Numerical Derivatives and Nonlinear Analysis

MATHEMATICAL CONCEPTS AND METHODS IN SCIENCE AND ENGINEERING

Series Editor: Angelo Miele
Mechanical Engineering and Mathematical Sciences
Rice University

Recent volumes in this series:

22 **APPLICATIONS OF FUNCTIONAL ANALYSIS IN ENGINEERING** • *J. L. Nowinski*

23 **APPLIED PROBABILITY** • *Frank A. Haight*

24 **THE CALCULUS OF VARIATIONS AND OPTIMAL CONTROL: An Introduction** • *George Leitmann*

25 **CONTROL, IDENTIFICATION, AND INPUT OPTIMIZATION** • *Robert Kalaba and Karl Spingarn*

26 **PROBLEMS AND METHODS OF OPTIMAL STRUCTURAL DESIGN** • *N. V. Banichuk*

27 **REAL AND FUNCTIONAL ANALYSIS, Second Edition Part A: Real Analysis** • *A. Mukherjea and K. Pothoven*

28 **REAL AND FUNCTIONAL ANALYSIS, Second Edition Part B: Functional Analysis** • *A. Mukherjea and K. Pothoven*

29 **AN INTRODUCTION TO PROBABILITY THEORY WITH STATISTICAL APPLICATIONS** • *Michael A. Golberg*

30 **MULTIPLE-CRITERIA DECISION MAKING: Concepts, Techniques, and Extensions** • *Po-Lung Yu*

31 **NUMERICAL DERIVATIVES AND NONLINEAR ANALYSIS** • *Harriet Kagiwada, Robert Kalaba, Nima Rasakhoo, and Karl Spingarn*

32 **PRINCIPLES OF ENGINEERING MECHANICS Volume 1: Kinematics—The Geometry of Motion** • *Millard F. Beatty, Jr.*

33 **PRINCIPLES OF ENGINEERING MECHANICS Volume 2: Dynamics—The Analysis of Motion** • *Millard F. Beatty, Jr.*

34 **STRUCTURAL OPTIMIZATION Volume 1: Optimality Criteria** • *Edited by M. Save and W. Prager*

Numerical Derivatives and Nonlinear Analysis

Harriet Kagiwada
Robert Kalaba
Nima Rasakhoo
and
Karl Spingarn

Hughes Aircraft Company
El Segundo, California

PLENUM PRESS • NEW YORK AND LONDON

Library of Congress Cataloging in Publication Data

Numerical derivatives and nonlinear analysis.

(Mathematical concepts and methods in science and engineering; 31)
Bibliography: p.
Includes index.
1. Numerical analysis—Data processing. I. Kagiwada, Harriet H., date. II. Series.
QA297.N854 1986 519.4 86-5062
ISBN 0-306-42178-X

A Division of Plenum Publishing Corporation
233 Spring Street, New York, N.Y. 10013

Printed in the United States of America

Preface

For many years it has been an article of faith of numerical analysts that the evaluation of derivatives of complicated functions should be avoided. Derivatives were evaluated using finite differences or, more recently, using symbolic manipulation packages. The first has the disadvantage of limited accuracy. The second has disadvantages of being expensive and requiring considerable computer memory.

The recent developments described in this text allow the evaluation of derivatives using simple automatic derivative evaluation subroutines programmed in FORTRAN or BASIC. These subroutines can even be programmed on a personal computer. The concept for the evaluation of the derivatives was originally developed by Wengert over 20 years ago. Significant improvements have been made in Wengert's method and are utilized in this text.

The purpose of this text is to familiarize computer users with a simple and practical method for obtaining the partial derivatives of complicated mathematical expressions. The text illustrates the use of automatic derivative evaluation subroutines to solve a wide range of nonlinear least-squares, optimal control, system identification, two-point boundary value problems, and integral equations. The numerical values of the derivatives are evaluated exactly, except for roundoff, using simple FORTRAN or BASIC subroutines. These derivatives are derived automatically behind the scenes, from the equivalent of analytical expressions, without any effort from the user. The use of costly software packages is not required. In many cases, such as in optimization problems, the user of the program need only enter an analytical expression into the program. The derivatives as well as the complete solution are then evaluated automatically.

Throughout this work our primary goal has been to emphasize the role of the computer in evaluating higher-order partial derivatives and what this implies as far as moving from a linear world view to one that is nonlinear. The last five chapter take up a variety of nonlinear structures with which modern applied mathematics deals and which should become of increasing importance in the future. We have not hesitated to repeat what we consider necessary to make each chapter more or less self-contained; and, where applicable, we have tried to sketch the derivations of the nonlinear equations to be dealt with.

The first chapter provides an introduction to the computer evaluation of partial derivatives. Wengert's basic approach is sketched first. This is followed by a description of the table or FEED (for fast and efficient evaluation of derivatives) method. Then a brief description of the important technique of Wexler is given. Finally, some applications to finding a root of a nonlinear equation using the Newton–Raphson and Halley methods are given.

Chapter 2 deals with nonlinear least squares. Here we wish to fit a nonlinear model containing some parameters to observations. The Newton–Raphson method is used, and a useful FEED program is discussed in detail. Then we show how to modify the program to handle a related problem. The chapter closes with a brief treatment of constrained optimization problems.

Chapter 3 provides an introduction to optimal control theory and the calculus of variations. Then approaches to the solution of such problems are explained, and finally programs for the automatic solution of these variational problems are presented and discussed.

System identification is the subject of Chapter 4. Here we wish to estimate constants in nonlinear ordinary differential equations so as to give the best fit to observations. The basic method used is quasilinearization, and a general program is given for automating the solution of this class of problems.

Nonlinear two-point boundary value problems are considered in Chapter 5. Recently, A. A. Sukhanov presented a new approach to these problems, and the role of machine evaluated partial derivatives is critical. Again, actual FORTRAN programs are provided and explained.

Finally, Chapter 6 treats the numerical solution of nonlinear integral equations. The approach through imbedding is presented, and the automatic solution is given in the form of a computer program in the BASIC language.

The text is designed to supplement upper division undergraduate and graduate texts used in existing courses on least-squares and regression analyses, optimal control, system identification, boundary value problems, computer science, etc. In fact, it can be used as a supplementary text for almost any course where computer solutions are desired with a minimum of time and effort. It provides students with user-friendly hands-on experience for rapidly obtaining computer solutions for a large number of homework assignments. Computer homework problems, which were previously assigned as a term project, can now be completed in less than one week. The text can also be used for self-study by practicing engineers and scientists to help solve real-world problems.

Over 20 years ago, R. Wengert (Reference 1) showed how automatic differentiation of complicated functions could be programmed. His essential idea was to consider all variables as functions of a hypothetical variable so that first and higher order partial derivatives could be evaluated "behind the scenes" while a function is being evaluated. The human supplied the knowledge of calculus, and the computer did the arithmetic. This caused quite a stir at the Rand Corporation, where Richard Bellman, Harriet Kagiwada, and Robert Kalaba, among others, were working on problems in orbit determination via quasilinearization (Reference 2). Wengert's method was tested extensively and found to be most useful (Reference 3), though certain drawbacks were noted.

To read the book, only a basic knowledge of differential calculus and FORTRAN is needed. For applications in the later chapters, though, a knowledge of the numerical solution of systems of ordinary differential equations, such as all engineering and economics students would have, is needed.

It will be seen that once the essential ideas for computer evaluation of partial derivatives are mastered, it is not difficult or time consuming to prepare required special purpose FEED libraries to deal with a situation at hand. FORTRAN or BASIC is more than adequate, and even an IBM PC has had sufficient computer power for certain problems.

Over the years, we have discussed numerical differentiation with many students and colleagues. Our deepest thanks go to R. Bellman, A. Tishler, A. Wexler, and especially to L. Tesfatsion.

The role of machine differentiation will grow, especially as an experimental tool for the investigation of methods for nonlinear phenomena. Now, on to numerical derivatives!

H. Kagiwada
R. Kalaba
N. Rasakhoo
K. Spingarn

El Segundo, California

Contents

1

Methods for Numerical Differentiation

It was in 1964 that R. Wengert (Reference 1) published a paper that showed how the partial derivatives of a given function could be evaluated by a computing machine without the user's first having formed the analytical expressions for the desired derivatives. Furthermore, no use was made of symbol manipulation or finite difference approximations. A salient idea was that the computer was to do arithmetic, solely, and the user was to supply the needed "savvy" about differential calculus. In a companion paper (Reference 2) Wilkins discussed some practical considerations associated with the Wengert method. These two papers were studied intensely at the Rand Corporation for, at that time, there was great interest in orbit determination via quasilinearization (References 3 and 4). This resulted in a third early paper (Reference 3). A book was published in 1981 that described the situation to that date (Reference 5). A language known as PROSE also appeared.

As we shall describe below, we were generally pleased with Wengert's method, but it had two serious flaws. One was that evaluation of partial derivatives was slow and inefficient owing to the need to reevaluate the function itself each time a partial derivative was to be evaluated. In addition, for third and higher partial derivatives there was an annoying ambiguity in the exact procedure to use.

Both of these difficulties disappeared in an approach suggested in 1983 (Reference 6); then in 1985 A. Wexler (Reference 7) made an important conceptual improvement.

1.1. Wengert's Method (Reference 1)

Let

$$z = f(x, y),$$

and consider x and y to be functions of t so that

$$\dot{z} = f_x\dot{x} + f_y\dot{y}$$

according to the chain rule of differential calculus. It is clear that

$$\dot{z} = f_x(x, y),$$

if $\dot{x} = 1$ and $\dot{y} = 0$. On the other hand,

$$\dot{z} = f_y(x, y),$$

if $\dot{x} = 0$ and $\dot{y} = 1$. Wengert showed how we may actually evaluate $\dot{z}$.

Assume that

$$f(x, y) = x + \ln xy.$$

Let X, Y, and Z, be three vectors with components x, $\dot{x}$; y, $\dot{y}$; and z, $\dot{z}$. Now consider the FORTRAN subroutine named FUN:

```
SUBROUTINE FUN(X,Y,Z)
DIMENSION X(2),Y(2),Z(2),A(2),B(2)
CALL MULT(X,Y,A)
CALL LOGG(A,B)
CALL ADD(X,B,Z)
RETURN
END
```

Here are subroutines MULT, LOGG, and ADD:

```
SUBROUTINE MULT(X,Y,Z)
DIMENSION X(2),Y(2),Z(2)
Z(1) = X(1)*Y(1)
Z(2) = X(2)*Y(1)+X(1)*Y(2)
RETURN
END
```

```
SUBROUTINE LOGG(X,Y)
DIMENSION X(2),Y(2)
Y(1) = ALOG(X(1))
Y(2) = (X(1)**(-1))*X(2)
RETURN
END

SUBROUTINE ADD(X,Y,Z)
DIMENSION X(2),Y(2),Z(2)
Z(1) = X(1)+Y(1)
Z(2) = X(2)+Y(2)
RETURN
END
```

It is clear that whenever subroutine FUN is called with X and Y as inputs, the values returned in Z are z and $\dot{z}$. These subroutines illustrate two important features of the Wengert approach. Several basic subroutines dealing with functions of two variables (addition, subtraction, multiplication, and division) are to be available, as are some subroutines dealing with the common functions of one variable: logarithm, sine, and so on. Then the actual function under consideration is to be constructed from a string of CALL statements, as in FUN.

Finally, to get the partial derivatives we employ a subroutine called PD(X,Y,DER). It reads this way:

```
SUBROUTINE PD(X,Y,DER)
DIMENSION X(2),Y(2),DER(2)
X(2) = 1.0
Y(2) = 0.0
CALL FUN(X,Y,Z)
DER(1) = Z(2)
X(2) = 0.0
Y(2) = 1.0
CALL FUN(X,Y,Z)
DER(2) = Z(2)
RETURN
END
```

When the call to PD is complete, DER(1) contains the partial derivative of z with respect to x, and DER(2) contains the partial derivative of z

with respect to y, both evaluated for $x = X(1)$ and $y = Y(1)$. If we also wish the value of z for these values of x and y, in subroutine PD we may add a scalar memory location VAL to the argument list and insert the assignment statement

$$\text{VAL} = Z(1)$$

after one of the assignment statements for the two-dimensional array DER. Notice that the function is evaluated with each call to subroutine FUN, an inefficient thing to do.

1.2. FEED (Fast and Efficient Evaluation of Derivatives) (Reference 6)

Let us now show how we can evaluate required partial derivatives of z without having to reevaluate the function each time. Suppose we wish to evaluate z, z_x, z_y, and z_{xx} for given values of x and y. We make Table 1.1 (again having $z = x + \ln xy$).

The table just given has these characteristics. By looking at the first column, it is clear that the first column evaluates z. The second, third, and fourth entries in each row are obtained from the entry in the first column by differentiation. Notice also from the third row on that no quantity appears on the right until it has appeared on the left earlier in the table. This means that we can evaluate z, z_x, z_y, and z_{xx} [they will be in locations $Z(1)$, $Z(2)$, $Z(3)$, $Z(4)$] by calling prewritten subroutines that handle the table row by row.

Table 1.1. Table for Example

Variable	$\partial/\partial x$	$\partial/\partial y$	$\partial^2/\partial x^2$
$A = x$	$A_x = 1$	$A_y = 0$	$A_{xx} = 0$
$B = y$	$B_x = 0$	$B_y = 1$	$B_{xx} = 0$
$C = AB$	$C_x = A_xB + AB_x$	$C_y = A_yB + AB_y$	$C_{xx} = A_{xx}B + 2A_xB_x + AB_{xx}$
$D = \ln C$	$D_x = C^{-1}C_x$	$D_y = C^{-1}C_y$	$D_{xx} = -C^{-2}C_x^2 + C^{-1}C_{xx}$
$Z = A + D$	$Z_x = A_x + D_x$	$Z_y = A_y + D_y$	$Z_{xx} = A_{xx} + D_{xx}$

To handle the first two rows, we may use the "vectorizing" routines

```
SUBROUTINE LIN1(X,A)
DIMENSION A(4)
A(1) = X
A(2) = 1.0
A(3) = 0.0
A(4) = 0.0
RETURN
END
```

and

```
SUBROUTINE LIN2(Y,B)
DIMENSION B(4)
B(1) = Y
B(2) = 0.0
B(3) = 1.0
B(4) = 0.0
RETURN
END
```

For row 3 we write

```
SUBROUTINE MULT(A,B,C)
DIMENSION A(4),B(4),C(4)
C(1) = A(1)*B(1)
C(2) = A(2)*B(1)+A(1)*B(2)
C(3) = A(3)*B(1)+A(1)*B(3)
C(4) = A(4)*B(1)+2.0*A(2)*B(2)+A(1)*B(4)
RETURN
END
```

For row 4 we may use

```
SUBROUTINE LOGG(C,D)
DIMENSION C(4),D(4)
D(1) = ALOG(C(1))
D(2) = C(2)/C(1)
D(3) = C(3)/C(1)
D(4) = -(C(2)/C(1))**2+C(4)/C(1)
RETURN
END
```

Lastly, for row 5, we may write

```
      SUBROUTINE ADD(A,D,Z)
      DIMENSION A(4),D(4),Z(4)
      DO 37 I = 1,4
   37 Z(I) = A(I)+D(I)
      RETURN
      END
```

The implementation of the table is completed in this way:

```
      SUBROUTINE FUN(X,Y,Z)
      DIMENSION A(4),B(4),C(4),D(4),Z(4)
      CALL LIN1(X,A)
      CALL LIN2(Y,B)
      CALL MULT(A,B,C)
      CALL LOGG(C,D)
      CALL ADD(A,D,Z)
      RETURN
      END
```

When values are inserted in X and Y and a call to FUN is completed, the desired numerical values of z, z_x, z_y, and z_{xx} will be found in memory locations $Z(1)$, $Z(2)$, $Z(3)$, and $Z(4)$.

1.3. An implementation of the FEED Procedure in BASIC

Consider the function

$$z = x + \ln xy.$$

A BASIC program that calculates, for given values of x and y, z, z_x, z_y, and z_{xx} is given below. It is followed by the output produced on an IBM PC.

```
10 FOR X = 1 to 3
20 FOR Y = 1 to 3
30 GOSUB 70
40 NEXT Y
50 NEXT X
60 END
70 Z = X+LOG(X*Y)
80 GOSUB 150
```

```
 90 GOSUB 240
100 GOSUB 290
110 GOSUB 340
120 LPRINT X,Y,Z,1+1/X,1/Y,-1/X^2
125 LPRINT X,Y,ZZ(1),ZZ(2),ZZ(3),ZZ(4)
130 RETURN
140 END
150 A(1) = X
160 A(2) = 1
170 A(3) = 0
180 A(4) = 0

190 B(1) = Y
200 B(2) = 0
210 B(3) = 1
220 B(4) = 0
230 RETURN
240 C(1) = A(1)*B(1)
250 C(2) = A(2)*B(1)+A(1)*B(2)
260 C(3) = A(3)*B(1)+A(1)*B(3)
270 C(4) = A(4)*B(1)+2*A(2)*B(2)+A(1)*B(4)
280 RETURN
290 E = C(1)
294 PRINT E
295 D(1) = LOG (E)
300 D(2) = C(2)/C(1)
310 D(3) = C(3)/C(1)
320 D(4) = (C(4)*C(1)-C(2)*C(2))/(C(1)*(C(1))
330 RETURN
340 FOR I = 1 TO 4
350 ZZ(I) = A(I)+D(I)
355 NEXT I
360 RETURN
```

```
1    1    1           2           1           -1
1    1    1           2           1           -1
1    2    1.693147    2           .5          -1
1    2    1.693147    2           .5          -1
1    3    2.098612    2           .3333334    -1
1    3    2.098612    2           .3333334    -1
2    1    2.693147    1.5         1           -.25
2    1    2.693147    1.5         1           -.25
2    2    3.386294    1.5         .5          -.25
2    2    3.386294    1.5         .5          -.25
2    3    3.79176     1.5         .3333334    -.25
2    3    3.79176     1.5         .3333334    -.25
3    1    4.098612    1.333333    1           -.1111111
3    1    4.098612    1.333333    1           -.1111111
3    2    4.79176     1.333333    .5          -.1111111
3    2    4.79176     1.333333    .5          -.1111111
3    3    5.197225    1.333333    .3333334    -.1111111
3    3    5.197225    1.333333    .3333334    -.1111111
```

1.4. Wexler's Approach (Reference 7)

Recently an important advance in automatic derivative evaluation was proposed by A. Wexler of the USC School of Medicine. Its main advantage is that it permits the user to evaluate an expression and its partial derivatives by writing a FORTRAN expression that closely resembles the algebraic expression under consideration, rather than using a sequence of call statements as has been explained. To present the method, we shall use as an example

$$z = x1 \cdot x2 + \exp(x3).$$

Our objective is the calculation of z and its first partial derivatives with respect to $x1$, $x2$, and $x3$ with, for example,

$$x1 = 4.0,$$

$$x2 = 6.0,$$

$$x3 = 0.0.$$

A program which accomplishes this is listed below. In looking at the main program, several features stand out. There is a long one-dimensional array, DA, in which the values of $x1$, $x2$, and $x3$ are placed, as well as their derivatives. Each variable requires four locations in the array DA, one for its value, and three for the values of its first partial derivatives. The integer-valued array IP contains the values 1, 5, 9, 13, . . . , 397, which are the locations in which our four-dimensional structures begin. The essential item to notice is the call to subroutine EVAL. The first argument is an expression formed by referring to the integer-valued function subprograms IADD, IMULT, and IEXP.

Here is what IADD(IP1,IP2) accomplishes. The four-dimensional structures in DA beginning at IP1 and IP2 are combined in the proper manner following the rules of differential calculus. The results are put in the four-dimensional structure beginning at location M in the array DA. The integer M is, in effect, increased by four with each call to IADD, IMULT, or IEXP. With each call the value returned is the first position in DA of the results.

After the FORTRAN compiler has evaluated the first argument in EVAL, our answers will be in four consecutive positions in DA beginning with IND = IP(IROW−1). The value of the expression and its three first derivatives are then placed in locations VAL and DERIV(1), DERIV(2), and DERIV(3).

The expression

IADD(IMULT(IS1,IS2),IEXP(IS3))

is so easy to interpret that its use represents a sizable gain over employing a sequence of CALL statements. By using concepts from dynamic allocation of memory, it is possible to cut down dramatically on the dimension of the work space array DA. Some output follows the program.

```
C          MAIN PROGRAM FOR NEW FEED
           COMMON DA(1000),IP(100), IROW
           DIMENSION DERIV(3)
C
C          FIRST DO THE EQUIVALENT OF LIN
C
           X1 = 4.0E+00
           X2 = 6.0E+00
           X3 = 0.0E+00
           DO 3  I= 1,100
      3   IP(I) = 4*I-3
          DA(1) = X1
          DA(2) = 1.0
          DA(3) = 0.0
          DA(4) = 0.0
          DA(5) = X2
          DA(6) = 0.0
          DA(7) = 1.0E+00
          DA(8) = 0.0E+00
          DA(9) = X3
          DA(10) = 0.0E+00
          DA(11) = 0.0E+00
          DA(12) = 1.0E+00
          IS1 = IP(1)
          IS2 = IP(2)
          IS3 = IP(3)
          IROW = 4
C
C         END ON LIN
C
          CALL EVAL(IADD(IMULT(IS1,IS2),IEXP(IS3)),VAL,DERIV)

          WRITE(11,100) VAL,(DERIV(I),I = 1,3)
    100   FORMAT(1X,E20.5)
          STOP 'END OF MY PROGRAM'
          END
C
C
          FUNCTION IADD(IP1,IP2)
          COMMON DA(1000),IP(100),IROW
```

```
      M = IP(IROW)
      DA(M) = DA(IP1)+DA(IP2)
      DO 15 I = 1,3
   15 DA(M+I) = DA(IP1+I)+DA(IP2+I)
      IADD = M
      IROW = IROW+1
      RETURN
      END
C
C
      FUNCTION IMULT(IP1,IP2)
      COMMON DA(1000),IP(100),IROW
      M = IP(IROW)
      DA(M) = DA(IP1)*DA(IP2)
      DO 15  I= 1,3
   15 DA(M+I) = DA(IP1)*DA(IP2+I)+DA(IP1+I)*DA(IP2)
      IMULT = M
      IROW = IROW+1
      RETURN
      END

C
C
      FUNCTION IEXP(IP1)
      COMMON DA(1000),IP(100),IROW
      M = IP(IROW)
      Z = DA(IP1)
      DA(M) = EXP(Z)
      DO 15 I = 1,3
   15 DA(M+I) = DA(MA)*DA(IP1+I)
      IROW = IROW+1
      IEXP = M
      RETURN
      END

C
C
      SUBROUTINE EVAL(II,VAL,DERIV)
      COMMON DA(1000),IP(100),IROW
      DIMENSION DERIV(3)
      IR1 = IROW-1
      IND = IP(IR1)
      VAL = DA(IND)
      DO 15 I = 1,3
   15 DERIV(I) = DA(IND+I)
      RETURN
      END

 0.25000E+02
 0.60000E+01
 0.40000E+01
 0.10000E+01
```

1.5. Higher-Order Methods for Finding Roots

Let us consider solving the equation

$$f(x) = 0.$$

We shall discuss the Newton and Halley procedures. One way of deriving Newton's method is to introduce the variable u as

$$u = f(x)$$

and then approximate x as a function of u, an inverse function, in a neighborhood of the current approximation x_0, by means of the linear expression

$$x = a + bu.$$

To obtain a and b, we enforce the above equation and the one formed by differentiating both sides with respect to x; i. e.,

$$x' = 1 = bu'.$$

The variables x, u, and u' are to be evaluated at $x = x_0$. Then for a and b we obtain

$$a = x - (1/u')u$$

and

$$b = (1/u').$$

We wish to know the value of x when $u = 0$, so the next approximation of the root is

$$x_1 = a = x_0 - [u(x_0)/u'(x_0)].$$

As we see, getting the next approximation involves evaluating $u(x_0)$ and $u'(x_0)$. This is the general step in the Newton–Raphson sequential approximation method, a most valuable procedure.

If we are able to evaluate more derivatives conveniently, then other possibilities are opened up. Consider, e. g., the equation

$$\arctan x = 0,$$

with a known root $x = 0$. Write

$$\arctan x = u$$

or

$$x = \tan u.$$

In passing from a current approximation of the root, x_0, to a new one, we could use the linear approximation given earlier. However, we see that x as a function of u has a pole at $u = \pi/2$. Let us then use a rational approximation,

$$x = (a + bu)/(1 + cu).$$

In this case, there are three constants to be determined, a, b, and c. We form the three equations

$$\begin{aligned} x + cux &= a \cdot 1 + bu, \\ 1 + c(ux)' &= a \cdot 0 + bu', \\ 0 + c(ux)'' &= a \cdot 0 + bu'', \end{aligned}$$

where the variables are to be evaluated at $x = x_0$. We need the values of x, ux, 1, and u and their first two derivatives with respect to x at $x = x_0$. Having these, and assuming that the resulting linear algebraic system is nonsingular, we can solve for the coefficients a, b, and c. Since we wish to know x when $u = 0$, the next approximation of the root is simply

$$x_1 = a.$$

The solution of the system above for a is

$$a = x - [u/u']/[1 - (uu''/2u'^2)],$$

where the right-hand side is evaluated at $x = x_0$, which is known as Halley's method.

When this method is applied to the equation $\arctan x = 0$, some pleasant surprises are in store for us. Starting with $x_0 = 1.0$, the next approximations are 0.12, 5.6×10^{-4}, and 6.5×10^{-11}. Starting with $x_0 = 1000$, the next approximations are 0.63, 5.1×10^{-2}, 4.4×10^{-5}. By contrast, Newton's method does not converge even for $x_0 = 10$, though it does for $x_0 = 0.1$.

For the initial approximations 0.1, 1, 10, 100, and 1000 we observe this behavior: When the approximation is far from the root at zero, c has a value such that $-1/c$, the position of the pole in the approximation, is close to $\pi/2$. When the approximation is close to zero, a and c are close to zero, and b is close to unity. This indicates that performance which is superior to Newton's method, with regard to number of iterations and domain of attraction, is achieved through the method's proper positioning

of the pole initially and through selecting the proper slope (the pole moving off to infinity) as the approximations to the root approach zero.

This illustrates the utility of having easy access to the values of first and second derivatives.

Exercises

1. Using the software described above for the table method, write a program that will solve the equation $x + \ln x = 0$ using Newton's method. As an initial approximation, use $x_0 = 0.5$ and obtain subsequent approximations from

$$x_{k+1} = x_k - f(x_k)/f'(x_k), \qquad k = 0, 1, 2, \ldots,$$

with $f(x) = x + \ln x$.

2. Solve the problem above using the Wengert and Wexler approaches to the calculation of the derivatives.

3. Derive the formula for Halley's method given earlier, and program the FEED procedure to solve $e^{-x} - x = 0$. Compare Halley's and Newton's method with regard to speed of convergence. (Answer: $x = 0.567143290$.)

4. For a function of ten independent variables, show that there are ten distinct first partial derivatives, 55 distinct second partial derivatives, and 220 distinct third partial derivatives.

2

Nonlinear Least Squares

One of the most common problems encountered in practical data analysis involves the fitting of a theoretical model to experimental data. Frequently the model takes the form of a dependent variable expressed as a function of several independent variables. Often the model will contain one or more parameters which have to be estimated. This estimation takes place on the basis of fitting the model to observations using the least-squares concept.

Here we shall focus our attention on models of the form

$$Q = f(K, L; \alpha, \beta), \tag{2.1}$$

where Q is viewed as a function of two independent variables, K and L, and α and β are two constants. We shall imagine that in a series of experiments of duration T, in the tth experiment K and L have the values K_t and L_t, and the observed value of Q is Q_t. Our objective is to estimate the values of the parameters α and β, which we shall do by minimizing the sum of squared errors, S,

$$S = S(\alpha, \beta) = \sum_{t=1}^{T} [f(K_t, L_t; \alpha, \beta) - Q_t]^2 \tag{2.2}$$

over the allowable values of α and β. The first-order necessary conditions for a minimum are, of course,

$$S_\alpha(\alpha, \beta) = 0, \tag{2.3}$$

$$S_\beta(\alpha, \beta) = 0. \tag{2.4}$$

As usual, $S_\alpha = \partial S/\partial\alpha$ and $S_\beta = \partial S/\partial\beta$. Generally speaking, these equations will constitute a system of two nonlinear equations in two unknowns. Among the many possible approaches to their solution let us use that of Newton and Raphson, which is a successive approximation scheme. Letting the current approximations be α_0 and β_0, we obtain the next approximation, α_1 and β_1, by solving the linearized equations

$$0 = S_\alpha(\alpha_0, \beta_0) + (\alpha_1 - \alpha_0)S_{\alpha\alpha}(\alpha_0, \beta_0) + (\beta_1 - \beta_0)S_{\alpha\beta}(\alpha_0, \beta_0), \tag{2.5}$$

$$0 = S_\alpha(\alpha_0, \beta_0) + (\alpha_1 - \alpha_0)S_{\beta\alpha}(\alpha_0, \beta_0) + (\beta_1 - \beta_0)S_{\beta\beta}(\alpha_0, \beta_0), \tag{2.6}$$

obtained by truncating the power series expansions of S_α and S_β after the linear terms in α and β. After having obtained α_1 and β_1, the equations above are used again to calculate the next approximations α_2 and β_2, etc.

From the definition of S in equation (2.2), we see that

$$S_\alpha = 2 \sum_{t=1}^{T} [f(K_t, L_t; \alpha, \beta) - Q_t]f_\alpha, \tag{2.7}$$

$$S_\beta = 2 \sum_{t=1}^{T} [f(K_t, L_t; \alpha, \beta) - Q_t]f_\beta. \tag{2.8}$$

For the second derivatives of S we have

$$S_{\alpha\alpha} = \sum_{t=1}^{T} \{f_\alpha^2(K_t, L_t; \alpha, \beta) + [f(K_t, L_t; \alpha, \beta) - Q_t]f_{\alpha\alpha}\}, \tag{2.9}$$

$$S_{\alpha\beta} = 2 \sum_{t=1}^{T} [f_\beta f_\alpha + (f - Q_t)f_{\alpha\beta}], \tag{2.10}$$

$$S_{\beta\beta} = 2 \sum_{t=1}^{T} [f_\beta^2 + (f - Q_t)f_{\beta\beta}]. \tag{2.11}$$

This shows that in forming the linearized equations we need, for $\alpha = \alpha_0$ and $\beta = \beta_0$, the values for the first partial derivations of f, f_α and f_β, and the values of the second partial derivatives of f, $f_{\alpha\alpha}$, $f_{\alpha\beta}$, and $f_{\beta\beta}$. In other words, for $\alpha = \alpha_0$ and $\beta = \beta_0$ we need the values of the gradient vector and Hessian matrix of the function f. As we know, these values are readily made available using the notions discussed earlier in Chapter 1.

We shall now describe a particular problem in economics and then look in detail at a FORTRAN program for its resolution. Then we shall use the same program, with small modifications, to resolve a problem in guidance and navigation, the problem of "passive ranging."

2.1. Fitting the CES Production Function

The CES production function in economic theory is given by

$$Q = f(K, L; \alpha, \beta) = [\alpha K^{\beta} + (1 - \alpha)L^{\beta}]^{1/\beta}, \tag{2.12}$$

in which K and L are capital and labor inputs, Q is the theoretical output, and α and β are certain constants that have to be estimated. Assume now that we have observations on K and L, the tth values being K_t and L_t, and we also have the corresponding observation on Q, Q_t, $t = 1, 2, \ldots, T$. We wish to find α and β that minimize S in equation (2.2). We shall use the Newton–Raphson successive approximation scheme to solve equations (2.3) and (2.4), and we shall use the FEED procedure to calculate the needed first and second partial derivatives of the function f given in equation (2.12).

To see how onerous this task is without FEED, the reader is invited to obtain, in analytical form, $f_{\beta\beta}$ from equation (2.12).

In this problem f is a function of two independent variables, α and β, and we wish, for given α and β, to determine f, f_{α}, f_{β}, $f_{\alpha\alpha}$, $f_{\alpha\beta}$, and $f_{\beta\beta}$, a total of six quantities. Let us now examine the FORTRAN program in Appendix A that accomplishes this. It consists of a main program and 19 simple subroutines, 15 of which comprise a useful little FEED library.

We begin with subroutines LINA and LINB. The first independent variable is α, and LINA accepts α as an input. It puts α in A(1), $\partial\alpha/\partial\alpha = 1$ in A(2), $\partial\alpha/\partial\beta = 0$ in A(3), $\partial^2\alpha/\partial\alpha^2 = 0$ in A(4), $\partial^2\alpha/\partial\alpha\partial\beta = 0$ in A(5), and $\partial^2\alpha/\partial\beta^2 = 0$ in A(6). Notice that the array A is six-dimensional, and this calculation and all others are done in double precision. The scalar α has been "vectorized."

The subroutine LINB accomplishes the same thing for the independent variable β. Here B(1) is assigned the value of β, and B(3) is assigned the value unity (corresponding to $\partial\beta/\partial\beta = 1$), zeros being placed in the remaining positions in the array B.

The subroutines NULL and UNIT create the constant functions identically zero and unity and their first and second partial derivatives with respect to α and β.

Consider the logarithmic function of one variable

$$z = \ln u, \tag{2.13}$$

where u is a function of α and β. Then we have, from calculus,

$$
\begin{aligned}
z &= \ln u, \\
z_\alpha &= (1/u)u_\alpha, \\
z_\beta &= (1/u)u_\beta, \\
z_{\alpha\alpha} &= -u^{-2}u_\alpha^{\ 2} + (1/u)u_{\alpha\alpha}, \\
z_{\alpha\beta} &= -u^{-2}u_\beta u_\alpha + (1/u)u_{\alpha\beta}, \\
z_{\beta\beta} &= -u^{-2}u_\beta^{\ 2} + (1/u)u_{\beta\beta}.
\end{aligned}
\tag{2.14}
$$

It would be possible to write subroutine LOGG treating u, u_α, u_β, $u_{\alpha\alpha}$, $u_{\alpha\beta}$, and $u_{\beta\beta}$ as inputs and z, z_α, z_β, $z_{\alpha\alpha}$, $z_{\alpha\beta}$, and $z_{\beta\beta}$ as outputs. We have found it to be more convenient, though, to consider a general function of one variable,

$$r = f(u), \tag{2.15}$$

where u is a function of α and β. Then,

$$
\begin{aligned}
r_\alpha &= f'(u)u_\alpha, \\
r_\beta &= f'(u)u_\beta, \\
r_{\alpha\alpha} &= f''(u)u_\alpha^{\ 2} + f'(u)u_{\alpha\alpha}, \\
r_{\alpha\beta} &= f''(u)u_\beta u_\alpha + f'(u)u_{\alpha\beta}, \\
r_{\beta\beta} &= f''(u)u_\beta^{\ 2} + f'(u)u_{\beta\beta}.
\end{aligned}
\tag{2.16}
$$

In these formulas we consider the values of u, u_α, u_β, $u_{\alpha\alpha}$, $u_{\alpha\beta}$, and $u_{\beta\beta}$ as inputs. Then for a given value of u, instructions for evaluating $f(u)$, $f'(u)$, and $f''(u)$ are to be given. And finally the values of r, r_α, r_β, $r_{\alpha\alpha}$, $r_{\alpha\beta}$, and $r_{\beta\beta}$ are to be output. Thus subroutine LOGG works this way. The values of u and its derivatives reside in U(1)–U(6). The values of $\ln u$, $1/u$, and $-1/u^2$ are put in memory locations F(1), F(2), F(3). Then subroutine DER is called.

Subroutine DER implements the calculus formulas in equations (2.16). The input values u, u_α, u_β, $u_{\alpha\alpha}$, $u_{\alpha\beta}$, and $u_{\beta\beta}$ are in U(1)–U(6). The values of $f(u)$, $f'(u)$, and $f''(u)$ are in F(1), F(2), and F(3). The output values are placed in Z(1)–Z(6).

It is now easy to see that the nonlinear functions of one variable

$$z = 1/u, \tag{2.17}$$

$$z = \ln u, \tag{2.18}$$

$$z = c^u, \tag{2.19}$$

$$z = e^u, \tag{2.20}$$

are handled by subroutines RECP, LOGG, CONTTU, and EXPP.

Subroutines ADD and SUB clearly handle addition and subtraction of two variables. Subroutine MULT(X,Y,Z) implements the calculus rules

$$
\begin{aligned}
z &= xy, \\
z_\alpha &= x_\alpha y + x y_\alpha, \\
z_\beta &= x_\beta y + x y_\beta, \\
z_{\alpha\alpha} &= x_{\alpha\alpha} y + 2x_\alpha y_\alpha + x y_{\alpha\alpha}, \\
z_{\alpha\beta} &= x_{\alpha\beta} y + x_\alpha y_\beta + x_\beta y_\alpha + x y_{\alpha\beta}, \\
z_{\beta\beta} &= x_{\beta\beta} y + 2x_\beta y_\beta + x y_{\beta\beta}.
\end{aligned}
\tag{2.21}
$$

Subroutine DIV(X,Y,Z) handles

$$z = x/y, \tag{2.22}$$

where x and y are functions of α and β. It does this by first calling RECP(Y,U) and then calling MULT(X,U,Z).

The function of two variables

$$z = u^v, \tag{2.23}$$

where u and v are both functions of α and β, is handled by subroutine UTTV(U,V,Z). This is done by using the sequence

$$
\begin{aligned}
a &= \ln u, \\
c &= av, \\
z &= e^c.
\end{aligned}
\tag{2.24}
$$

As the fifteenth and final FEED routine we mention subroutine MULCON. It merely multiplies a variable and its derivatives by a constant; this situation occurs so often that we have found this routine quite useful in practice.

Now let us consider the CES production function given in equation (2.12). For given values of the variables α and β and for given values of the constants K and L, this function, its two first and three second partial derivatives with respect to α and β are evaluated in subroutine CES(ALFA, BETA,AK,AL,Q). Basically all that this subroutine does is call the subroutines in the FEED library in the proper order and store the intermediate results.

The first column of the table is

$$
\begin{aligned}
A &= \alpha, \\
B &= \beta, \\
C &= K^B, \\
D &= AC, \\
R &= 1, \\
E &= R - A, \\
F &= L^B, \\
G &= EF, \\
H &= D + G, \\
P &= 1/B, \\
Q &= H^P.
\end{aligned} \tag{2.25}
$$

When a call to CES is complete, Q(1), . . . , Q(6) contain the values of the function in equation (2.12), its two first derivatives, and its three second derivatives with respect to α and β.

It is the essential simplicity of this type of subroutine that accounts for the ease with which the derivatives of a function may be evaluated.

Let us now look at subroutine GEDAT, the function of which is to generate the data. Corresponding values of K, L, and Q are put in corresponding positions in the one-dimensional arrays AKK, ALL, and AQQ. In our test case

$$
\begin{gathered}
\alpha = 0.4, \\
\beta = -0.7, \\
K_1 = 1, \qquad L_1 = 1, \\
\left.\begin{aligned} K_{i+1} &= K_i + 0.25i \\ L_{i+1} &= L_i + 0.1i^2 \end{aligned}\right\} \qquad i = 1, 2, \ldots, N-1,
\end{gathered} \tag{2.26}
$$

where $N = T$ is the number of observations. Finally,

$$
Q_i = [\alpha K_i^{\beta} + (1 - \alpha)L_i^{\beta}]^{1/\beta}, \qquad i = 1, 2, \ldots, N. \tag{2.27}
$$

Now we are ready to look at subroutine SSE. In the do-loop ending at the statement numbered 1, the sum in equation (2.2) is formed. Then the values of S and its first and second partial derivatives with respect to α and β, all for given values of α and β, are made available in the six-dimensional array ASS2.

Table 2.1. Results of Estimation, $T = 100$

Iteration No.	Estimated parameter		F_α	F_β	F
	α	β			
0	0.3	−0.6	-0.14×10^5	0.46×10^7	0.68×10^6
1	0.3262335	−0.7238237	-0.43×10^7	0.14×10^7	0.12×10^6
2	0.3516337	−0.7809449	-0.10×10^7	0.33×10^6	0.11×10^5
5	0.3992093	−0.7011998	-0.15×10^5	0.51×10^4	3.02
7	0.3999999	−0.7000001	−0.26	0.08	0.93×10^{-9}
9	0.4000000	−0.7000000	-0.14×10^{-8}	0.46×10^{-9}	0.15×10^{-24}

Subroutine NEWT simply solves the linear algebraic equations in equations (2.5) and (2.6). The routine receives the current values of α and β and the values of S, S_α, S_β, $S_{\alpha\alpha}$, $S_{\alpha\beta}$, and $S_{\beta\beta}$ in the six-dimensional vector named SS2. The linear algebraic equations, equations (2.5) and (2.6), are actually solved for the differences $\alpha_1 - \alpha_0$ and $\beta_1 - \beta_0$ (in memory locations Z1 and Z2). Finally, α_0 and β_0 are updated by Z1 and Z2, so that when a call to NEWT is completed, the new values of α and β are available in ALFA and BETA. It is assumed, of course, that the system of linear algebraic equations is nonsingular and not ill conditioned.

Having the subroutines just described, we are now ready to look at the main program. First N is set equal to 100 to indicate the number of observations, and NIT is set equal to 25 to indicate the number of Newton–Raphson iterations. Then GEDAT is called to generate the observations. ALFA is initialized to 0.3, and BETA is initialized to −0.6 (recall that the true values are assigned in GEDAT as 0.4 and −0.7). Then in the do-loop the call to SSE puts the current values of S and its derivatives in the array SS2. Then NEWT uses these values to update α and β until the iterative process is completed.

Results of a calculation are given in Table 2.1.

2.2. Passive Ranging

Once a program such as the one just presented is available, it is important to see how it is modified to handle similar problems. Let us now consider estimating the position of a stationary beacon on the basis of angular measurements. In the horizontal (x, y) plane a beacon is located at the fixed position (α, β). An observer moves along the x axis, and, when at position x_i, measures the angle (in radians) between the positive direction of the x axis and a ray directed toward the beacon. The observed value of this angle is denoted θ_i; its true value is $\arctan [\beta/(\alpha - x_i)]$. We assume that

$$\theta_i = \arctan [\beta/(\alpha - x_i)] + \varepsilon_i, \qquad i = 1, 2, \ldots, N, \tag{2.28}$$

where ε_i is a small "random" error. We are to estimate the position of the beacon by minimizing the sum S,

$$S = \sum_{i=1}^{N} \{\theta_i - \arctan [\beta/(\alpha - x_i)]\}^2. \tag{2.29}$$

We see that the problem is of the type just treated. Our main concern is with the programmatic changes that are induced. Parenthetically, though, we remark that the problem is of practical importance, and there are difficulties in its solution. Consider, e. g., that the beacon is on the x axis; each true angle is zero, no matter the range to the beacon. Now refer to Appendix B.

Since the nonlinear function arctan appears, we first augment our FEED library by adding this nonlinear function to it. The subroutine AATT does this (see Appendix B). It uses the calculus formulas

$$\begin{aligned} z &= \arctan u, \\ dz/du &= (1 + u^2)^{-1}, \\ d^2z/du^2 &= -(1 + u^2)^{-2}(2u), \end{aligned} \tag{2.30}$$

the values of z and its derivatives being put in the one-dimensional array F. Then a call to subroutine DER puts the values z, z_α, z_β, $z_{\alpha\alpha}$, $z_{\alpha\beta}$, and $z_{\beta\beta}$ in the array Z.

No changes are required in subroutine SSE, but subroutine CES now has to calculate the values of $\arctan [\beta/(\alpha - x_i)]$ and its first and second partial derivatives with respect to α and β.

Subroutine NEWT requires no changes. Subroutine GEDAT has to produce the noisy observations. The true values of the position coordinates are $\alpha = 20.0$ and $\beta = 1.0$. Observations are made with the observer at 0.1, 0.2, 0.3, ..., 10.0, a total of 100 observations. Each true value of an angular observation has noise added to it, noise which takes the form of a sample from a normal distribution with zero mean and standard deviation of 1.0 milliradian.

Finally, we consider the main program. The number of observations, N, is 100, and the number of iterations, NIT, to be done is 8. (This is a working program, and, no doubt, the first assignment, NIT = 25, was left over from earlier runs.) The initial estimates, $\alpha = 20.0$ and $\beta = 1.0$ (which are the true values), are inserted; note that the values of α and β which minimize S in equation (2.28) in general will not be these true values owing to the noise that is added to the true angular observations. The little do-loop calls for the Newton–Raphson iterations to be performed. Successive values of both α and β are printed in subroutine NEWT, and some output is provided.

The output in Appendix C for iteration number five shows that the minimum of S is achieved at $\alpha = 20.0375$ and $\beta = 1.0034$; and the minimum value of S is approximately 1×10^{-5}. The values of the S_α and S_β are about 10^{-18}. The determinant of the Hessian matrix of S, though not zero, is very small, about 0.2×10^{-7}.

2.3. Constrained Optimization

In optimization problems it often happens that a degree of realism can be added to the modeling effort by constraining the decision variables. Let us take a look at a general class of such problems and see how the automatic calculation of first and second partial derivatives bears on the situation.

Suppose that we wish to minimize the function $f(x_1, x_2, \ldots, x_n)$, subject to the m constraints

$$g^i(x_1, x_2, \ldots, x_n) = 0, \qquad i = 1, 2, 3, \ldots, m < n. \tag{2.31}$$

We introduce the Lagrange multipliers, $\lambda_1, \lambda_2, \ldots, \lambda_m$, and then find the first-order necessary conditions to be

$$\frac{\partial f}{\partial x_j} = \sum_{i=1}^{m} \lambda_i \frac{\partial g^i}{\partial x_j}, \qquad j = 1, 2, \ldots, n, \tag{2.32}$$

together with the m equations of constraint given above. In all, there are $(m + n)$ unknowns: $\lambda_1, \lambda_2, \ldots, \lambda_m; x_1, x_2, \ldots, x_n$. Our task is to find them by solving the equations above. Note that in merely forming these equations the number of first partial derivatives occurring is $(m + 1)n$; for f, a function of ten variables subject to six constraints, the number of first partial derivatives is 70. If the functions f and $g^1, g^2, \ldots, g^m$ are of any complexity, this would be an unpleasant prospect, both as far as forming the partial derivatives and programming them are concerned.

To reduce the size of the system of unknowns, we may consider solving, in principle, the first m equations in equation (2.32) for $\lambda_1, \lambda_2, \ldots, \lambda_m$; for, in these variables, the equations are linear and, we shall assume, nonsingular. Then we may substitute these values for $\lambda_1, \lambda_2, \ldots, \lambda_m$ into the remaining n-m equations in equation (2.32). In this way, the Lagrange multipliers are eliminated, and we are left with a system of n equations for $x_1, x_2, \ldots, x_n$, the first m of which are

$$\begin{aligned} g^1(x_1, x_2, \ldots, x_n) &= 0, \\ &\ldots \\ g^m(x_1, x_2, \ldots, x_n) &= 0. \end{aligned} \tag{2.33}$$

To represent the others, we need to introduce some notation. Let λ be an m-dimensional column vector whose ith component is λ_i. Let F_1 be an m-dimensional column vector given by

$$F_1 = (\partial f/\partial x_1, \ldots, \partial f/\partial x_m)^T, \tag{2.34}$$

where, as usual, the superscript T indicates transposition. The $m \times m$ matrix G_1 is defined by

$$G_1 = (\partial g^j/\partial x_i). \tag{2.35}$$

Thus, in vector-matrix notation, the first m relations in equation (2.32) may be represented as

$$G_1 \lambda = F_1 \tag{2.36}$$

and, in view of the assumed nonsingularity of the matrix G_1,

$$\lambda = G_1^{-1} F_1. \tag{2.37}$$

To represent the remaining $n - m$ equations in equation (2.32), we introduce the $(n - m)$-dimensional column vector F_2,

$$F_2 = (\partial f/\partial x_{m+1}, \ldots, \partial f/\partial x_n)^T, \tag{2.38}$$

and the matrix G_2, of dimension $(n - m) \times m$,

$$\begin{aligned} G_2 &= (\partial g^j/\partial x_{m+i}), \\ i &= 1, 2, \ldots, n - m, \\ j &= 1, 2, \ldots, m, \end{aligned} \tag{2.39}$$

so that

$$G_2\lambda = F_2 \tag{2.40}$$

completes the system. Upon the elimination of λ, the system of n equations for the n unknowns $x_1, x_2, \ldots, x_n$ becomes

$$0 = g \qquad (m \text{ equations}), \tag{2.41}$$

$$0 = G_2 G_1^{-1} F_1 - F_2 \qquad (n - m \text{ equations}), \tag{2.42}$$

where g is the m-dimensional column vector whose ith element is g^i.

If we elect to use the Newton–Raphson method of solution, it will be necessary to evaluate not only the right-hand sides of the equations, but also the partial derivatives of each component of the right-hand sides with respect to $x_1, x_2, \ldots, x_n$; i.e., second partial derivatives of the given functions $f, g, \ldots, g^m$ will have to be evaluated at the current approximations to the optimizing values of the decision variables, $x_1, x_2, \ldots, x_n$. As far as the first m equations above are concerned, this presents no problem. What is to be done, though, about the components of the right-hand sides of the second set?

We have

$$\frac{\partial}{\partial x_j}[G_2 G^{-1} F_1 - F_2] = \frac{\partial G_2}{\partial x_j} G_1^{-1} F_1 + G_2 \frac{\partial G_1^{-1}}{\partial x_j} F_1 + G_2 G_1^{-1} \frac{\partial F_1}{\partial x_j} - \frac{\partial F_2}{\partial x_j}, \tag{2.43}$$

The only term that presents any difficulty in its evaluation is $\partial G_1^{-1}/\partial x_j$. However, this too disappears when we recall that

$$\frac{\partial G_1^{-1}}{\partial x_j} = -G_1^{-1} \frac{\partial G_1}{\partial x_j} G_1^{-1}. \tag{2.44}$$

Thus by being able to evaluate the functions $f, g^1, \ldots, g^m$ at the current approximation to the minimizing values of the decision variables $x_1, x_2, \ldots, x_n$, as well as their first-order and distinct second-order partial derivatives, use of the Newton–Raphson method becomes possible. In all, there are $(m + 1)$ functions, $(m + 1)n$ first partial derivatives, and $(m + 1)n \times (n + 1)/2$ distinct second-order partial derivatives.

If the function f is a function of ten variables which are subject to six constraints, there are 70 first partial derivatives to be evaluated, as we have remarked, and $(6 + 1)\ 10 \cdot 11/2 = 385$ distinct second-order partial derivatives to be evaluated. In our FEED subroutines we shall have to dimension the vectors at $1 + 10 + 55 = 66$. In view of the virtually unlimited memories of modern computers and their high speeds of calculation ($\sim 10^7$ double precision multiplications per second), these requirements are modest. We also see that these problems lend themselves well to parallel computation, a task for the future. Some numerical results are quoted in Reference 3, but much remains to be done with regard to the exploitation of optimization methods involving the use of higher-order partial derivatives.

Appendix A: Nonlinear Least Squares Using FEED

```
C     MAIN PROGRAM FOR ECONOMETRIC PROBLEM
      IMPLICIT REAL *8(A-H,O-Z)
      DIMENSION AKK(100),ALL(100),AQQ(100),SS2(6)
      N=100
      NIT=25
      CALL GEDAT(N,AKK,ALL,AQQ)
      ALFA=0.3D0
      BETA=-0.6D0
      DO 1 IJK=1,NIT
      WRITE(6,11)IJK
      CALL SSE(N,AKK,ALL,AQQ,ALFA,BETA,SS2)
      CALL NEWT(ALFA,BETA,SS2)
1     CONTINUE
11    FORMAT(2X,//,2X'ITERATION NO.',I4,2X,//)
      STOP
      END
      SUBROUTINE SSE(N,AKK,ALL,AQQ,ALFA,BETA,ASS2)
      IMPLICIT REAL *8(A-H,O-Z)
      DIMENSION AKK(100), ALL(100), AQQ(100), QQQ(6), A7(6),
     $A8(6), A9(6)
     $B1(6), SS2(6),RR(6),ASS2(6)
      EON = 1.D0/DFLOAT(N)
      ONE = 1.D0
      CALL NULL (SS2)
      DO 1I=1,N
      AK=AKK(I)
      AL=ALL(I)
      CALL CES(ALFA,BETA,AK,AL,QQQ)
      AQ7=AQQ(I)
      CALL UNIT(A7)
      CALL MULCON(AQ7,A7,A8)
      CALL SUB(QQQ,A8,A9)
      CALL MULT(A9,A9,B1)
      CALL ADD(SS2,B1,RR)
      CALL MULCON(ONE,RR,SS2)
1     CONTINUE
      CALL MULCON(EON,SS2,ASS2)
      RETURN
      END
```

```
      SUBROUTINE GEDAT(N,AKK,ALL,AQQ)
      IMPLICIT REAL *8(A-H,O-Z)
      DIMENSION AKK(100),ALL(100),AQQ(100)
      ALFA=0.4D0
      BETA=-0.7D0
      AKK(1)=1.D0
      ALL(1)=1.D0
      N1=N-1
      DO 20 I=1,N1
      AII=DFLOAT(I)
      AKK(I+1)=AKK(I)+0.25D0*AII
      ALL(I+1)=ALL(I)+0.1D0*AII*AII
20    CONTINUE
      DO 21 I=1,N
      A1=BETA*DLOG(AKK(I))
      A2=BETA*DLOG(ALL(I))
      AAA=ALFA*DEXP(A1)+(1.D0-ALFA)*DEXP(A2)
      A3=(1.D0/BETA)*DLOG(AAA)
      AQQ(I)=DEXP(A3)
21    CONTINUE
      DO 22 I=1,N
22    WRITE(6,25)AKK(I),ALL(I),AQQ(I)
25    FORMAT (2X,3(D20.6,2X))
      RETURN
      END
      SUBROUTINE NEWT(ALFA,BETA,SS2)
      IMPLICIT REAL*8(A-H,O-Z)
      DIMENSION SS2(6)
      WRITE(6,62)
      WRITE(6,61)(SS2(I),I=1,6)
      WRITE(6,62)
      WRITE(6,63)SS2(4),SS2(5)
      WRITE(6,63)SS2(5),SS2(6)
      DET=SS2(4)*SS2(6)-SS2(5)*SS2(5)
      Z1=(SS2(5)*SS2(3)-SS2(6)*SS2(2))/DET
      Z2=(SS2(5)*SS2(2)-SS2(4)*SS2(3))/DET
      ALFA=Z1+ALFA
      BETA=Z2+BETA
      WRITE(6,62)
      WRITE(6,61)ALFA,BETA,Z1,Z2,DET
61    FORMAT(2X,6(D20.13,1X))
62    FORMAT(2X,////)
63    FORMAT(2X,4(D25.12,2X))
      RETURN
      END
      SUBROUTINE CES(ALFA,BETA,AK,AL,Q)
      IMPLICIT REAL*8(A-H,O-Q)
      DIMENSION
     $A(6),B(6),C(6),D(6),R(6),E(6),F(6),G(6),H(6),P(6),Q(6)
      CALL LINA(ALFA,A)
      CALL LINB(BETA,B)
      CALL CONTTU(AK,B,C,)
      CALL MULT(A,C,D)
      CALL UNIT(R)
      CALL SUB(R,A,E,)
      CALL CONTTU(AL,B,F)
```

```
      CALL MULT(E,F,G)
      CALL ADD(D,G,H)
      CALL RECP(B,P)
      CALL UTTV(H,P,Q)
      RETURN
      END
      SUBROUTINE DER(F,U,Z)
      IMPLICIT REAL*8(A-H,O-Z)
      DIMENSION F(3),Z(6),U(6)
      Z(1)=F(1)
      Z(2)=F(2)*U(2)
      Z(3)=F(2)*U(3)
      Z(4)=F(3)*U(2)*U(2)+F(2)*U(4)
      Z(5)=F(3)*U(3)*U(2)+F(2)*U(5)
      Z(6)=F(3)*U(3)*U(3)+F(2)*U(6)
      RETURN
      END
      SUBROUTINE LINA(ALFA,A)
      IMPLICIT REAL*8(A-H,O-Z)
      DIMENSION A(6)
      DO 1 I=1,6
    1 A(I)=0.D0
      A(1)=ALFA
      A(2)=1.D0
      RETURN
      END
      SUBROUTINE LINB(BETA,B)
      IMPLICIT REAL*8(A-H,O-Z)
      DIMENSION B(6)
      DO 1 I=1,6
    1 B(I)=0.D0
      B(1)=BETA
      B(3)=1.D0
      RETURN
      END
      SUBROUTINE RECP(X,Y)
      IMPLICIT REAL*8(A-H,O-Z)
      DIMENSION X(6),Y(6),F(3)
      F(1)=1.D0/X(1)
      F(2)=-1.D0/(X(1)*X(1))
      F(3)=2.D0/(X(1)*X(1)*X(1))
      CALL DER(F,X,Y)
      RETURN
      END
      SUBROUTINE LOGG(U,R)
      IMPLICIT REAL*8(A-H,O-Z)
      DIMENSION U(6),R(6),F(3)
      A=U(1)
      F(1)=DLOG(A)
      F(2)=1.D0/U(1)
      F(3)=-F(2)*F(2)
      CALL DER(F,U,R)
      RETURN
      END
      SUBROUTINE CONTTU(CON,U,Z)
      IMPLICIT REAL*8(A-H,O-Z)
```

```
      DIMENSION U(6),Z(6),F(3)
      A=U(1)
      B=DLOG(CON)
      A1=A*B
      F(1)=DEXP(A1)
      F(2)=F(1)*B
      F(3)=F(2)*B
      CALL DER(F,U,Z)
      RETURN
      END
      SUBROUTINE EXPP(V,C)
      IMPLICIT REAL*8(A-H,O-Z)
      DIMENSION V(6),C(6),F(3)
      A=V(1)
      F(1)=DEXP(A)
      F(2)=F(1)
      F(3)=F(1)
      CALL DER(F,V,C)
      RETURN
      END
      SUBROUTINE UNIT(R)
      IMPLICIT REAL*8(A-H,O-Z)
      DIMENSION R(6)
      DO 1I=1,6
    1 R(I)=0.D0
      R(1)=1.D0
      RETURN
      END
      SUBROUTINE NULL(A)
      IMPLICIT REAL*8(A-H,O-Z)
      DIMENSION A(6)
      DO 1I=1,6
    1 A(I)=0.D0
      RETURN
      END
      SUBROUTINE MULCON(C,A,B)
      IMPLICIT REAL*8(A-H,O-Z)
      DIMENSION A(6),B(6)
      DO 1I=1,6
      B(I)=C*A(I)
    1 CONTINUE
      RETURN
      END
      SUBROUTINE ADD(X,Y,Z)
      IMPLICIT REAL*8(A-H,O-Z)
      DIMENSION X(6),Y(6),Z(6)
      DO 1I=1,6
      Z(I)=X(I)+Y(I)
    1 CONTINUE
      RETURN
      END
      SUBROUTINE SUB(A,B,C)
      IMPLICIT REAL*8(A-H,O-Z)
      DIMENSION A(6),B(6),C(6)
      DO 1I=1,6
      C(I)=A(I)-B(I)
```

```
1     CONTINUE
      RETURN
      END
      SUBROUTINE MULT(X,Y,Z)
      IMPLICIT REAL*8(A-H,O-Z)
      DIMENSION X(6),Y(6),Z(6)
      Z(1)=X(1)*Y(1)
      Z(2)=X(2)*Y(1)+Y(2)*X(1)
      Z(3)=X(3)*Y(1)+Y(3)*X(1)
      Z(4)=X(4)*Y(1)+Y(4)*X(1)+2.DO*X(2)*Y(2)
      Z(5)=X(5)*Y(1)+Y(3)*X(2)+Y(5)*X(1)*+X(3)*Y(2)
      Z(6)=X(6)*Y(1)+Y(6)*X(1)+2.DO*X(3)*Y(3)
      RETURN
      END
      SUBROUTINE DIV(X,Y,Z)
      IMPLICIT REAL*8(A-H,O-Z)
      DIMENSION X(6),Y(6),Z(6),U(6)
      CALL RECP(Y,U)
      CALL MULT(X,U,Z)
      RETURN
      END
      SUBROUTINE UTTV(U,V,Z)
      IMPLICIT REAL*8(A-H,O-Z)
      DIMENSION U(6),V(6),Z(6),A(6),C(6)
      CALL LOGG(U,A)
      CALL MULT(A,V,C)
      CALL EXPP(C,Z)
      RETURN
      END
```

Appendix B: Modifications for Passive Ranging Problem

```
C       MAIN PROGRAM FOR PASSIVE RANGING
        IMPLICIT REAL*8(A-H,O-Z)
        DIMENSION AKK(100),ALL(100),AQQ(100),SS2(6)
        N=10
        N=100
        NIT=25
        NIT=8
        CALL GEDAT(N,AKK,ALL,AQQ)
        ALFA=11.5D+00
        BETA=.95D+00
         ALFA=11.2D+00
        BETA=.98D+00
        ALFA=11.3D+00
        BETA=.97D+00
        ALFA=11.01D+00
        BETA=1.01D+00
        ALFA=11.0D+00
        BETA=1.0D+00
        ALFA=20.0D+00
        DO 1 IJK=1,NIT
        WRITE(6,11) IJK
        CALL SSE(N,AKK,ALL,AQQ,ALFA,BETA,SS2)
        CALL NEWT(ALFA,BETA,SS2)
```

```
1       CONTINUE
11      FORMAT(2X,//,2X,'ITERATION.NO.,I4,2X,//)
        STOP
        END
        SUBROUTINE GEDAT(N,AKK,ALL,AQQ)
        IMPLICIT REAL*8(A-H,O-Z)
        REAL R
        DIMENSION AKK(100),ALL(100),AQQ(100)
        ALFA=11.0D+00
        ALFA=20.0D+00
        BETA=1.0D+00
        DEL=1.0D+00/DFLOAT(N)
        DEL=.1D+00
        DO 20 I=1,N
        AKK(I)=DFLOAT(I)
        AKK(I)=DEL*DFLOAT(I)
        ALL (I)=0.0D+00
 20     CONTINUE
        DO 21 I=1,N
        TANG=(BETA-ALL(1))/(ALFA-AKK(I))
        THETA=DATAN(TANG)
        STD=.00001D+00
        STD=.001D+00
        R=GNORM(0)
        RR=DBLE(R)
        THETA=THETA+RR*STD
        AQQ(I)=THETA
 21     CONTINUE
        DO 22 I=1,N
 22     WRITE(6,25) AKK(I),ALL(I),AQQ(I)
 25     FORMAT(1X,3D25.6)
        RETURN
        END
        SUBROUTINE CES(ALFA,BETA,AK,AL,Q)
        IMPLICIT REAL*8(A-H,O-Z)
        DIMENSION A(6),B(6),C(6),D(6),E(6),F(6),G(6),H(6),Q(6)
        CALL LINA(ALFA,A)
        CALL LINB(BETA,B)
        CALL UNIT(C)
        CALL MULCON(AL,C,D)
        CALL SUB(B,D,E)
        CALL MULCON(AK,C,F)
        CALL SUB(A,F,G)
        CALL DIV(E,G,H)
        CALL AATT(H,Q)
        RETURN
        END
        SUBROUTINE AATT(U,Z)
        IMPLICIT REAL*8(A-H,O-Z)
        DIMENSION U(6) ,Z(6),F(3)
        F(1)=DATAN(U(1))
        F(2)=1.0D+00/(1.0D+00+U(1)*U(1))
        F(3)=-2.0D+00*U(1)*F(2)*F(2)
        CALL DER(F,U,Z)
        RETURN
        END
```

Appendix C: Sample Output for Passive Ranging

```
0.100000D 00    0.0    0.509242D-01
0.200000D 00    0.0    0.503688D-01
0.300000D 00    0.0    0.498224D-01
0.400000D 00    0.0    0.509103D-01
0.500000D 00    0.0    0.526440D-01
0.600000D 00    0.0    0.517567D-01
0.700000D 00    0.0    0.533776D-01
0.800000D 00    0.0    0.537283D-01
0.900000D 00    0.0    0.508316D-01
0.100000D 01    0.0    0.514620D-01
0.110000D 01    0.0    0.535007D-01
0.120000D 01    0.0    0.540250D-01
0.130000D 01    0.0    0.556421D-01
0.140000D 01    0.0    0.528626D-01
0.150000D 01    0.0    0.550696D-01
0.160000D 01    0.0    0.560994D-01
0.170000D 01    0.0    0.555981D-01
0.180000D 01    0.0    0.534829D-01
0.190000D 01    0.0    0.562936D-01
0.200000D 01    0.0    0.577300D-01
0.210000D 01    0.0    0.552211D-01
0.220000D 01    0.0    0.570497D-01
0.230000D 01    0.0    0.565308D-01
0.240000D 01    0.0    0.559584D-01
0.250000D 01    0.0    0.571373D-01
0.260000D 01    0.0    0.576457D-01
0.270000D 01    0.0    0.572209D-01
0.280000D 01    0.0    0.573212D-01
0.290000D 01    0.0    0.580797D-01
0.300000D 01    0.0    0.582929D-01
0.310000D 01    0.0    0.585768D-01
0.320000D 01    0.0    0.594716D-01
0.330000D 01    0.0    0.598058D-01
0.340000D 01    0.0    0.611447D-01
0.350000D 01    0.0    0.606481D-01
0.360000D 01    0.0    0.611292D-01
0.370000D 01    0.0    0.620502D-01
0.380000D 01    0.0    0.606796D-01
0.390000D 01    0.0    0.625840D-01
0.400000D 01    0.0    0.606326D-01
0.410000D 01    0.0    0.621598D-01
0.420000D 01    0.0    0.626719D-01
0.430000D 01    0.0    0.625491D-01
0.440000D 01    0.0    0.630679D-01
0.450000D 01    0.0    0.643489D-01
0.460000D 01    0.0    0.651552D-01
0.470000D 01    0.0    0.635337D-01
0.480000D 01    0.0    0.649839D-01
0.490000D 01    0.0    0.661724D-01
0.500000D 01    0.0    0.664009D-01
0.510000D 01    0.0    0.664785D-01
0.520000D 01    0.0    0.679577D-01
0.530000D 01    0.0    0.685368D-01
```

```
0.540000D 01          0.0          0.674148D-01
0.550000D 01          0.0          0.684010D-01
0.560000D 01          0.0          0.697783D-01
0.570000D 01          0.0          0.713133D-01
0.580000D 01          0.0          0.719369D-01
0.590000D 01          0.0          0.707553D-01
0.600000D 01          0.0          0.694972D-01
0.610000D 01          0.0          0.722092D-01
0.620000D 01          0.0          0.744015D-01
0.630000D 01          0.0          0.714410D-01
0.640000D 01          0.0          0.736447D-01
0.650000D 01          0.0          0.739556D-01
0.660000D 01          0.0          0.758123D-01
0.670000D 01          0.0          0.759639D-01
0.680000D 01          0.0          0.741672D-01
0.690000D 01          0.0          0.762474D-01
0.700000D 01          0.0          0.752540D-01
0.710000D 01          0.0          0.775557D-01
0.720000D 01          0.0          0.759946D-01
0.730000D 01          0.0          0.770164D-01
0.740000D 01          0.0          0.790764D-01
0.750000D 01          0.0          0.799849D-01
0.760000D 01          0.0          0.805726D-01
0.770000D 01          0.0          0.825394D-01
0.780000D 01          0.0          0.822684D-01
0.790000D 01          0.0          0.824028D-01
0.800000D 01          0.0          0.845082D-01
0.810000D 01          0.0          0.824504D-01
0.820000D 01          0.0          0.843215D-01
0.830000D 01          0.0          0.864161D-01
0.840000D 01          0.0          0.871383D-01
0.850000D 01          0.0          0.879189D-01
0.860000D 01          0.0          0.856591D-01
0.870000D 01          0.0          0.872341D-01
0.880000D 01          0.0          0.900648D-01
0.890000D 01          0.0          0.902476D-01
0.900000D 01          0.0          0.896046D-01
0.910000D 01          0.0          0.930997D-01
0.920000D 01          0.0          0.912601D-01
0.930000D 01          0.0          0.949478D-01
0.940000D 01          0.0          0.937859D-01
0.950000D 01          0.0          0.945652D-01
0.960000D 01          0.0          0.963808D-01
0.970000D 01          0.0          0.967459D-01
0.980000D 01          0.0          0.982089D-01
0.990000D 01          0.0          0.997891D-01
0.100000D 01          0.0          0.979554D-01

ITERATION NO.  1

0.1073754313449D-05  0.3573444441031D-06  -0.5790767233128D-05
0.5844035707876D-04  -0.7489736890556D-03  0.9957968541996D-02

    0.584403570788D-04          -0.748973689056D-03
   -0.748973689056D-03           0.995796854200D-02
```

```
0.2003710666795D 02  0.1003372443393D 01  0.3710666795489-01
0.3372443393272D-02  0.2098565047579D-07

ITERATION NO.  2

0.1070605642745D-05  0.1269461537022D-08 -0.2648211541986D-07
0.5817974904362D-04 -0.7453543191347D-03  0.9902467128032D-02

      0.581797490436D-04             -0.745354319135D-03
     -0.745354319135D-03              0.990246712803D-02

0.1003745512498D 02  0.1003401345894D 01  0.3484570235740D-03
0.2890250068726D-04  0.2056999136892D-07

ITERATION NO.  3

0.1070605481215D-05 0.1215012773958D-13 -0.8845701552967D-12
0.5817667645869D-04 -0.7453164343351D-03  0.9901949167750D-02

      0.581766764587D-04             -0.745316434335D-03
     -0.745316434335D-03              0.990194916775D-02

0.2003745515119D 02  0.1003401347956D 01  0.2620719618623D-07
0.2061939909492D-08  0.2056590575260D-07

ITERATION NO.  4

0.1070605481215D-05 -0.6183340514956D-19  0.7329206686002D-18
0.5817667620194D-04 -0.7453164313230D-03  0.9901949128835D-02

      0.581766762019D-04             -0.745316431323D-03
     -0.745316431323D-03              0.990194912883D-02

0.2003745515119D 02  0.1003401347956D 01  0.3209847253943D-14
0.1675863216635D-15  0.2056590543633D-07

ITERATION NO.  5

0.1070605481215D-05 -0.6183340514956D-19  0.7329206686002D-18
0.5817667620194D-04 -0.7453164313230D-03  0.9901949128835D-02

      0.581766762019D-04             -0.745316431323D-03
     -0.745316431323D-03              0.990194912883D-02

0.2003745515119D 02  0.1003401347956D 01  0.3209847253943D-14
0.1675863216635D-15  0.2056590543633D-07

ITERATION NO.  6

0.1070605481215D-05 -0.6183340514956D-19  0.7329206686002D-18
0.5817667620194D-04 -0.7453164313230D-03  0.9901949128835D-02

  0.581766762019D-04                 -0.745316431323D-03
 -0.745316431323D-03                  0.990194912883D-02
```

3

Optimal Control

The automatic solution of a certain class of optimal control problems is described in this chapter. Optimal control problems involve the solution of two-point boundary value problems. The derivatives in the two-point boundary value equations are evaluated automatically and the complete solution of the optimal control problem is obtained. None of the derivatives usually associated with the Euler–Lagrange equations, Pontryagin's maximum principle, the Newton–Raphson method, or the gradient method need be calculated by hand. Depending on the problem and the method of solution used, the user of the program need only specify the initial conditions and the terminal time, and input one of the following by calling the appropriate FORTRAN subroutines: (i) the integrand of the cost functional and the differential constraints if applicable, (ii) the integrand of the cost functional and the Hamiltonian function, or (iii) only the Hamiltonian function.

In Section 3.1 the derivations of the two-point boundary value problem equations for solving the optimal control problems are given using (i) the Euler–Lagrange equations and (ii) Pontryagin's maximum principle. The equations for the numerical solution of the two-point boundary value problem are given in Section 3.2 using (i) the Newton–Raphson method and (ii) the gradient method. The functions, the order of the derivatives, and the number of the derivatives required for the automatic solution are also described in Section 3.2. A description of the subroutines is given in Section 3.3 and examples of optimal control problems are given in Section 3.4. The program listings are given in Section 3.5.

3.1. Control Theory

The derivations of the equations used in optimal control theory are given in this section. The Euler–Lagrange equations are derived for the simplest problem in the calculus of variations (References 1–10) and for optimal control problems. The Hamiltonian function is introduced and the equations are derived for solving optimal control problems using Pontryagin's maximum principle (References 1–6).

3.1.1. Euler–Lagrange Equations

The simplest problem in the calculus of variations involves the extremization of an integral, called a functional, in one independent variable and an unknown dependent function. To obtain the extremum, the unknown function must satisfy the Euler–Lagrange equations. These equations are derived as follows.

Consider the problem of determining a function $y = y(t)$, $a \leq t \leq b$, which satisfies the boundary conditions

$$y(a) = y_0, \qquad y(b) = y_1 \tag{3.1}$$

and which minimizes the integral J,

$$J = \int_a^b F(t, y, \dot{y})\, dt, \tag{3.2}$$

where F is a given function of its three arguments, and

$$\dot{y} = dy/dt. \tag{3.3}$$

This problem is called the simplest problem in the calculus of variations. Assume that the function that minimizes J is

$$y(t) = x(t), \qquad a \leq t \leq b \tag{3.4}$$

and consider a nearby function

$$y(t) = x(t) + \varepsilon\eta(t), \tag{3.5}$$

where ε is a small parameter, positive or negative. The function $\eta = \eta(t)$, $a \leq t \leq b$, is a variation function which is arbitrary, except that

$$\eta(a) = \eta(b) = 0, \tag{3.6}$$

since only curves that fulfill the boundary conditions are admissible. Use of this nonoptimal curve will not lower the value of the integral J, since $x = x(t)$ is the minimizing function, so that

$$\int_a^b F(t, x, \dot{x})\, dt \leq \int_a^b F(t, x + \varepsilon\eta, \dot{x} + \varepsilon\dot{\eta})\, dt. \tag{3.7}$$

Expansion of the right-hand side of the above inequality using Taylor's theorem then yields

$$\int_a^b F(t, x, \dot{x})\, dt \leq \int_a^b F(t, x, \dot{x})\, dt + \varepsilon \int_a^b [\eta(t)F_x(t, x, \dot{x}) + \dot{\eta}(t)F_{\dot{x}}(t, x, \dot{x})]\, dt + o(\varepsilon) \tag{3.8}$$

where

$$F_x = \frac{\partial F}{\partial x}, \qquad F_{\dot{x}} = \frac{\partial F}{\partial \dot{x}} \tag{3.9}$$

and the term $o(\varepsilon)$ refers to terms involving ε to the second and higher powers. Subtracting we obtain

$$0 \leq \varepsilon \int_a^b [\eta(t)F_x(t, x, \dot{x}) + \dot{\eta}F_{\dot{x}}(t, x, \dot{x})]\, dt + o(\varepsilon). \tag{3.10}$$

Now for $|\varepsilon|$ sufficiently small, the sign of the right-hand side of the above relation is determined by the first term, so that we must have

$$0 \leq \varepsilon \int_a^b [\eta(t)F_x(t, x, \dot{x}) + \dot{\eta}F_{\dot{x}}(t, x, \dot{x})]\, dt. \tag{3.11}$$

But the only way that this relation can hold for all sufficiently small values of ε, both positive and negative, is that

$$0 = \int_a^b [\eta(t)F_x(t, x, \dot{x}) + \dot{\eta}(t)F_{\dot{x}}(t, x, \dot{x})]\, dt. \tag{3.12}$$

We now wish to make use of the arbitrariness of the function $\eta = \eta(t)$. Through integration by parts we find that

$$0 = \int_a^b \eta(t)\left[F_x(t, x, \dot{x}) - \frac{d}{dt} F_{\dot{x}}(t, x, \dot{x})\right] dt + \eta(t)F_{\dot{x}}(t, x, \dot{x})\Big|_a^b \tag{3.13}$$

But we have $\eta(a) = \eta(b) = 0$, so that the boundary terms disappear, and we are left with the equation

$$0 = \int_a^b \eta(t)\left[F_x(t, x, \dot{x}) - \frac{d}{dt} F_{\dot{x}}(t, x, \dot{x})\right] dt. \tag{3.14}$$

Since this relation must hold for arbitrary choice of the function $\eta = \eta(t)$, we must have

$$F_x(t, x, \dot{x}) - \frac{d}{dt} F_{\dot{x}}(t, x, \dot{x}) = 0, \qquad a \leq t \leq b, \tag{3.15}$$

which is called the Euler–Lagrange equation for the simplest problem in the calculus of variations.

The Euler–Lagrange equation was derived above for fixed initial and terminal boundary conditions using the calculus of variations approach. In optimal control, the initial condition, $x(0)$, is often fixed with the terminal condition, $x(T)$, unspecified. The functional

$$J = \int_0^T F(t, x, \dot{x})\, dt \tag{3.16}$$

with boundary conditions

$$x(0) = c, \qquad x(T) = \text{unspecified}, \tag{3.17}$$

is then minimized by the solution of the Euler–Lagrange equation

$$F_x(t, x, \dot{x}) - \frac{d}{dt} F_{\dot{x}}(t, x, \dot{x}) = 0 \tag{3.18}$$

with associated initial and transversality conditions

$$x(0) = c, \qquad F_{\dot{x}}(t, x, \dot{x})\Big|_{t=T} = 0. \tag{3.19}$$

The transversality condition at the terminal boundary $t = T = b$ is obtained from equation (3.13) since $\eta(T)$ is arbitrary when $x(T)$ is unspecified.

The above equations are two-point boundary value problems since the boundary conditions must be satisfied at both the initial and the terminal ends.

3.1.2. Pontryagin's Maximum Principle

Consider the optimal control problem (References 3, 4) of extremizing the cost functional

$$J = \int_0^T F(t, \mathbf{x}, y)\, dt \tag{3.20}$$

subject to the nonlinear differential constraints and initial conditions

$$\dot{\mathbf{x}} = \mathbf{f}(t, \mathbf{x}, y), \qquad \mathbf{x}(0) = \mathbf{x}_0, \tag{3.21}$$

where $\mathbf{x}$ is an n-dimensional state vector, y is the scalar control, and $\mathbf{f}$ is an n-dimensional vector the components of which, $f_1, f_2, \ldots, f_n$, are nonlinear function of $\mathbf{x}$ and y. Adjoin equation (3.21) to equation (3.20) with the multiplier $\mathbf{p}(t)$. Then

$$J = \int_0^T \{F(t, \mathbf{x}, y) + \mathbf{p}^T[f(t, \mathbf{x}, y) - \dot{\mathbf{x}}]\}\, dt. \tag{3.22}$$

Define the Hamiltonian function

$$H = F(t, \mathbf{x}, y) + \mathbf{p}^T f(t, \mathbf{x}, y) \tag{3.23}$$

Substituting equation (3.23) into (3.22) and integrating the last term by parts we obtain

$$J = \mathbf{p}^T(0)\mathbf{x}(0) - \mathbf{p}^T(T)\mathbf{x}(T) + \int_0^T (H + \dot{\mathbf{p}}^T\mathbf{x})\, dt \tag{3.24}$$

The variation in J due to the variation in y is then

$$\Delta J = \mathbf{p}^T(0)\, \Delta\mathbf{x}(0) - \mathbf{p}^T(T)\, \Delta\mathbf{x}(T) + \int_0^T \left\{\left[\left(\frac{\partial H}{\partial \mathbf{x}}\right)^T + \dot{\mathbf{p}}^T\right] \Delta\mathbf{x} + \frac{\partial H}{\partial y}\, \Delta y\right\} dt \tag{3.25}$$

To obtain an extremum, ΔJ must be zero. This is obtained for

$$\dot{\mathbf{p}} = -\frac{\partial H}{\partial \mathbf{x}}, \tag{3.26}$$

$$\frac{\partial H}{\partial y} = 0 \tag{3.27}$$

and the boundary condition

$$\mathbf{p}(T) = 0. \tag{3.28}$$

The initial condition is fixed so that

$$\Delta\mathbf{x}(0) = 0. \tag{3.29}$$

Equations (3.21) and (3.26)–(3.28) determine the optimal control and state vector using Pontryagin's maximum principle. The numerical solution of these two-point boundary value problems is discussed in the next section.

3.2. Numerical Methods

Nonlinear optimal control problems involve the solution of nonlinear two-point boundary value problems. The numerical solution of these problems using the automatic derivative evaluation method is discussed in this section. For nonlinear problems, iterative methods must be used to obtain the numerical solution. Two iterative methods are discussed, (i) the Newton–Raphson method and (ii) the gradient method.

The Newton–Raphson method generally yields more accurate numerical results than the gradient method, but requires the use of higher-order derivatives and requires the control function to be solved explicitly as a function of the state and costate variables. The gradient method converges to the final numerical solution more slowly, but requires evaluation of only up to the first- or second-order derivatives, and does not require the control function to be solved for explicitly. The gradient method is thus more suitable for solving the general nth-order nonlinear system optimal control problem.

The automatic derivative evaluation method is the same as discussed earlier in this text and uses the table method, which is a modification of Wengert's original method (References 11, 12). The automatic derivative evaluation method is used here to obtain completely the automatic solution of optimal control problems (References 13–18). In this section the functions, the order of the derivatives, and the number of the derivatives which must be evaluated automatically are also described.

3.2.1. Newton–Raphson Method

The Newton–Raphson method for obtaining the numerical solution of optimal control problems can be used with either the Euler–Lagrange equations or the equations obtained using Pontryagin's maximum principle. As we shall see, however, the Euler–Lagrange equations, as used in this section, require the evaluation of derivatives higher than the second order.

3.2.1.1. Newton–Raphson Method Using the Euler–Lagrange Equations

A typical optimal control problem (References 1–10) consists of extremizing the cost functional

$$J = \int_0^T F_1(t, \mathbf{x}, y)\, dt \tag{3.30}$$

subject to the nonlinear differential constraints and initial conditions

$$\dot{\mathbf{x}} = \mathbf{f}(t, \mathbf{x}, y), \qquad \mathbf{x}(0) = \mathbf{x}_0 \tag{3.31}$$

where $\mathbf{x}$ is an n-dimensional state vector, y is the scalar control, and $\mathbf{f}$ is an n-dimensional vector the components of which, $f_1, f_2, \ldots, f_n$, are nonlinear functions of $\mathbf{x}$ and y. The vectors are defined by

$$\mathbf{x} = (x_1, x_2, \ldots, x_n)^T, \tag{3.32}$$

$$\mathbf{f} = (f_1, f_2, \ldots, f_n)^T. \tag{3.33}$$

Consider the one-dimensional case

$$\dot{x} = f(t, x, y), \qquad x(0) = c \tag{3.34}$$

where x is the scalar state, y is the scalar control, and f is a nonlinear function of x and y. A typical subset of the problems represented by equations (3.30) and (3.34) is the linear regulator for a first-order system when F_1 is quadratic and f is linear.

Solving equation (3.34) for y and substituting into equation (3.30) yields the calculus of variations problem for the extremization of the integral

$$J = \int_0^T F(t, x, \dot{x})\, dt \tag{3.35}$$

with the initial condition

$$x(0) = c, \tag{3.36}$$

where $x = x(t)$, $0 \leq t \leq T$, is an unknown function of the independent variable t. The integrand F is a given function of t, x, and $\dot{x}$, where

$$\dot{x} = dx/dt. \tag{3.37}$$

To obtain the extremization, the unknown function $x(t)$ must satisfy the

Euler–Lagrange equation

$$F_x - \frac{d}{dt} F_{\dot{x}} = 0 \tag{3.38}$$

with boundary conditions

$$x(0) = c; \qquad F_{\dot{x}}(T) = 0, \tag{3.39}$$

where

$$F_x = \frac{\partial f}{\partial x}, \qquad F_{\dot{x}} = \frac{\partial F}{\partial \dot{x}}, \qquad F_{\dot{x}}(T) = \left.\frac{\partial F}{\partial \dot{x}}\right|_{t=T} \tag{3.40}$$

Using the chain rule of differentiation, equation (3.38) can be expressed in the form

$$F_x - [F_{\dot{x}t} + F_{\dot{x}x}\dot{x} + F_{\dot{x}\dot{x}}\ddot{x}] = 0, \tag{3.41}$$

which is a second-order ordinary differential equation.

Newton–Raphson Equations for a First-Order System. The solution is obtained using the Newton–Raphson method (References 1, 19). Solving equation (3.41) for $\ddot{x}$, the optimizing function $x(t, s)$ is the solution of the initial value equation

$$\ddot{x} = \frac{F_x - F_{\dot{x}t} - F_{\dot{x}x}\dot{x}}{F_{\dot{x}\dot{x}}} \tag{3.42}$$

with initial conditions

$$x(0, s) = c; \qquad \dot{x}(0, s) = s, \tag{3.43}$$

where s is the unknown initial slope at $t = 0$. The Newton–Raphson equations are obtained by expanding the boundary condition

$$F_{\dot{x}}(T) = 0 \tag{3.44}$$

in a Taylor series and truncating after the first-order terms

$$F_{\dot{x}}(T) + (s_{k+1} - s_k)F_{\dot{x}s}(T) = 0, \tag{3.45}$$

where s_k is the current approximation of the initial slope s. Solving for s_{k+1} yields the recurrence relation

$$s_{k+1} = s_k - \frac{F_{\dot{x}}(T)}{F_{\dot{x}s}} \tag{3.46}$$

Equation (3.46) can be expressed in a form that is more useful for the automatic calculation of derivatives. Using the chain rule of differentiation

$$F_{\dot{x}s} = F_{\dot{x}x}x_s + F_{\dot{x}\dot{x}}\dot{x}_s, \tag{3.47}$$

where

$$F_{\dot{x}s} = \frac{\partial F_{\dot{x}}}{\partial s} \quad \text{and} \quad x_s = \frac{\partial x}{\partial s}. \tag{3.48}$$

Substituting equation (3.47) into (3.46) yields the equation

$$s_{k+1} = s_k - \left(\frac{F_{\dot{x}}}{F_{\dot{x}x}x_s + F_{\dot{x}\dot{x}}\dot{x}_s} \right)_{t=T} \tag{3.49}$$

Differentiating equation (3.42) with respect to s, we have

$$\begin{aligned} \ddot{x}_s = [F_{\dot{x}\dot{x}}(F_{xx}x_s + F_{x\dot{x}}\dot{x}_s - F_{\dot{x}tx}x_s - F_{\dot{x}t\dot{x}}\dot{x}_s - F_{\dot{x}xx}\dot{x}x_s - F_{\dot{x}x\dot{x}}\dot{x}\dot{x}_s \\ - F_{\dot{x}x}\dot{x}_s) - (F_x - F_{\dot{x}t} - F_{\dot{x}x}\dot{x})(F_{\dot{x}\dot{x}x}x_s + F_{\dot{x}\dot{x}\dot{x}}\dot{x}_s)]/F_{\dot{x}\dot{x}}^2 \end{aligned} \tag{3.50}$$

with initial conditions

$$x_s(0, s) = 0, \qquad \dot{x}_s(0, s) = 1. \tag{3.51}$$

Equations (3.42) and (3.50) can be expressed in terms of first-order differential equations by defining

$$x_1 = x, \qquad f_1 = \ddot{x}, \tag{3.52}$$

$$x_2 = \dot{x}, \qquad f_2 = \ddot{x}_s, \tag{3.53}$$

$$x_3 = x_s, \tag{3.54}$$

$$x_4 = \dot{x}_s, \tag{3.55}$$

where f_1 and f_2 are formed automatically in the program from the more elementary partial derivatives, F_x, F_{xx}, $F_{\dot{x}}$, $F_{\dot{x}\dot{x}}$, $F_{\dot{x}\dot{x}\dot{x}}$, etc. in equations (3.42) and (3.50). Then

$$\dot{x}_1 = x_2, \qquad x_1(0) = c \tag{3.56}$$

$$\dot{x}_2 = f_1, \qquad x_2(0) = s \tag{3.57}$$

$$\dot{x}_3 = x_4, \qquad x_3(0) = 0 \tag{3.58}$$

$$\dot{x}_4 = f_2, \qquad x_4(0) = 1 \tag{3.59}$$

The four initial value equations (3.56)–(3.59) are integrated from $t = 0$ to $t = T$ with the initial approximation $x_2(0) = s$. The new value of s is calculated from equation (3.49) and the initial value equations (3.56)–(3.59) are integrated again, etc., until there is little or no change in s.

Automatic Derivative Evaluation. The user of the program for the automatic solution of the above optimal control problem need only input the integrand, F, of the cost functional and specify the initial conditions and the terminal time.

Equation (3.50) shows that all the partial derivatives of F up to the third must be evaluated (References 13, 14), i. e., $F_{\dot{x}\dot{x}\dot{x}}$, $F_{\dot{x}\dot{x}x}$, $F_{\dot{x}xx}$, etc. For simplicity, express F as a function of its three arguments, t, x, and $\dot{x}$. Then

$$F = F(t, x, \dot{x}) \tag{3.60}$$

and the partial derivatives are

$$F_1 = \frac{\partial F}{\partial t}, \qquad F_2 = \frac{\partial F}{\partial x}, \qquad F_3 = \frac{\partial F}{\partial \dot{x}}, \tag{3.61}$$

$$F_{11} = \frac{\partial^2 F}{\partial t\, \partial t}, \qquad F_{12} = \frac{\partial^2 F}{\partial t\, \partial x}, \qquad \text{etc.} \tag{3.62}$$

Since the order of differentiation is immaterial, for example, $F_{21} = F_{12}$, etc., only six second derivatives need be calculated. In a similar manner the number of third derivatives that must be calculated is 10. The derivation of the number of second and third derivatives is given in the next section.

The function $F(t, x, \dot{x})$ and all of its derivatives are represented in the automatic solution program by the vector **F(I)**. The above discussion indicates that this vector must consist of 20 components, i. e., **I** = 1–20, as shown in Table 3.1. In order to compute the derivatives automatically the 20-component vector must be computed for each of the three variables, t, x, and $\dot{x}$, and for each of the functions of the three variables, such as the sum, the product, the square root, the quotient, etc. The subroutines required to compute these vectors are as follows: (1) linear, (2) constant, (3) add, (4) multiplication, (5) division, and (6) function. Each of these

Table 3.1. Number of Components of the Vector, **F(I)**, for a First-Order System, Newton–Raphson Method and Euler–Lagrange Equations

Function	Number of components
$F(t, x, \dot{x})$	1
First derivatives	3
Second derivatives	6
Third derivatives	10

Table 3.2. Definitions of the Vector Components for a First-Order System, Newton–Raphson Method and Euler–Lagrange Equations

Vector component L	Derivative arguments i	j	k	Symbol[a]
1	—			z
2	1			$\partial z/\partial t$
3	2			$\partial z/\partial x$
4	3			$\partial z/\partial \dot{x}$
5	1	1		$\partial^2 z/\partial t\, \partial t$
6	1	2		$\partial^2 z/\partial t\, \partial x$
7	1	3		$\partial^2 z/\partial t\, \partial \dot{x}$
8	2	2		$\partial^2 z/\partial x\, \partial x$
9	2	3		$\partial^2 z/\partial x\, \partial \dot{x}$
10	3	3		$\partial^2 z/\partial \dot{x}\, \partial \dot{x}$
11	1	1	1	$\partial^3 z/\partial t\, \partial t\, \partial t$
12	1	1	2	$\partial^3 z/\partial t\, \partial t\, \partial x$
13	1	1	3	$\partial^3 z/\partial t\, \partial t\, \partial \dot{x}$
14	1	2	2	$\partial^3 z/\partial t\, \partial x\, \partial x$
15	1	2	3	$\partial^3 z/\partial t\, \partial x\, \partial \dot{x}$
16	1	3	3	$\partial^3 z/\partial t\, \partial \dot{x}\, \partial \dot{x}$
17	2	2	2	$\partial^3 z/\partial x\, \partial x\, \partial x$
18	2	2	3	$\partial^3 z/\partial x\, \partial x\, \partial \dot{x}$
19	2	3	3	$\partial^3 z/\partial x\, \partial \dot{x}\, \partial \dot{x}$
20	3	3	3	$\partial^3 z/\partial \dot{x}\, \partial \dot{x}\, \partial \dot{x}$

[a] z is a scalar variable equal to t, x, or $\dot{x}$ or a function of the variables, such as the integrand F.

subroutines is discussed in more detail below and in Section 3.3. The computation and storage of the derivatives in the above manner is called the table method. The definitions of the vector components for the table method are given in Table 3.2.

Subroutine Linear. In subroutine LIN the variables t, x, and $\dot{x}$ are represented by $X1$, $X2$, and $X3$, respectively, and the corresponding vectors are represented by **A**, **B**, and **C** as shown in Table 3.3. The first derivatives are equal to unity, while all the other derivatives are equal to zero. From Table 3.2, we have

$$A(1) = t, \qquad B(1) = x, \qquad C(1) = \dot{x}, \tag{3.63}$$

$$A(2) = 1, \qquad B(3) = 1, \qquad C(4) = 1 \tag{3.64}$$

Table 3.3. FORTRAN Representation of the Variables for a First-Order System, Newton–Raphson Method and Euler–Lagrange Equations

Variables	FORTRAN program	
	Variables	Vectors
t	$X1$	**A**
x	$X2$	**B**
$\dot{x}$	$X3$	**C**

and all other components of the 20-component vectors **A**, **B**, and **C** are equal to zero.

Number of Second and Third Derivatives. The number of second and third derivatives in the vectors, **A**, **B**, **C**, **F**, etc. is determined as follows. Forming a matrix of all permutations of the second derivatives

$$\begin{bmatrix} F_{11} & F_{12} & F_{13} \\ F_{21} & F_{22} & F_{32} \\ F_{31} & F_{32} & F_{33} \end{bmatrix} \tag{3.65}$$

it is seen that because of symmetry only six of the nine second derivatives need be calculated, i.e., F_{11}, F_{12}, F_{13}, F_{22}, F_{23}, and F_{33}. The number of permutations of the second derivatives is nine while the number of the combinations is six. Note that the order of combinations, as shown in Table 3.2, is such that in forming the derivatives, F_{ij}, first i is set equal to 1 while j is incremented from 1 to 3. Once j reaches its maximum limit of 3, then i is incremented to 2 while j is incremented from i to 3. Finally i is incremented to 3 and j is set equal to i. This sequence for forming the second derivatives is used in the multiplication subroutine for forming the derivatives of the product of two variables.

The number of permutations of the rth derivatives of n variables is equal to the number of permutations of r elements selected with replacement from an n-set of distinguishable objects. The number of permutations of the third derivatives is $n^r = 3^3 = 27$. The number of combinations of the rth derivatives of n variables is equal to the number of combinations of $(n + r - 1)$ taken r at a time (Reference 20). The number of combinations of the third derivatives is given by

$$\binom{n + r - 1}{r} = \frac{(n + r - 1)!}{r!(n - 1)!} = \frac{(3 + 3 - 1)!}{3!(3 - 1)!} = 10. \tag{3.66}$$

The combinations of the third derivatives that must be calculated are

1. F_{111}	5. F_{123}	9. F_{233}
2. F_{112}	6. F_{133}	10. F_{333}
3. F_{113}	7. F_{222}	
4. F_{122}	8. F_{223}	

The order of the combinations is such that in forming the derivatives, F_{ijk}, first i and j are set equal to 1, while k is incremented from 1 to 3. Once k reaches its maximum limit of 3, then j is incremented to 2 while k is incremented from j to 3. When k again reaches its maximum limit of 3, then j is incremented to 3 while k is set equal to j. Next i is incremented to 2, while j is set equal to i, and k is incremented from j to 3. When k reaches its maximum limit, then j is incremented to 3, and k is set equal to 3. Finally, i is incremented to 3 while j and k are set equal to i.

3.2.1.2. Newton–Raphson Method Using Pontryagin's Maximum Principle

Consider now the case where the cost functional given by equation (3.30) is extremized using Pontryagin's maximum principle instead of the Euler–Lagrange equations. The Hamiltonian function is

$$H_1 = F_1(t, \mathbf{x}, y) + \mathbf{p}^T\mathbf{f}(t, \mathbf{x}, y), \tag{3.67}$$

where $\mathbf{p}(t)$ is the vector of the costate variables. The control $y(t)$ that minimizes H_1 is obtained from the equation

$$\frac{\partial H}{\partial y} = 0. \tag{3.68}$$

Solving equation (3.68) for y and substituting into equations (3.30) and (3.31) we obtain the cost functional

$$J = \int_0^T F(t, \mathbf{x}, \mathbf{p})\, dt \tag{3.69}$$

subject to the differential constraints

$$\dot{\mathbf{x}} = \mathbf{g}(t, \mathbf{x}, \mathbf{p}), \qquad \mathbf{x}(0) = \mathbf{x}_0. \tag{3.70}$$

For the two-dimensional case, equations (3.69) and (3.70) are functions of t, x_1, x_2, p_1, and p_2. The control, y, has been eliminated by substitution. The Hamiltonian function and differential constraints are

$$H = F(t, x_1, x_2, p_1, p_2) + p_1 g_1 + p_2 g_2, \tag{3.71}$$

$$\dot{x}_1 = g_1(t, x_1, x_2, p_1, p_2), \qquad x_1(0) = x_{10}, \tag{3.72}$$

$$\dot{x}_2 = g_2(t, x_1, x_2, p_1, p_2), \qquad x_2(0) = x_{20}. \tag{3.73}$$

The costate variables, p_1 and p_2, are the solutions of the differential equations and boundary conditions

$$\dot{p}_1 = -\frac{\partial H}{\partial x_1} = -[F_{x_1} + p_1 g_{1x_1} + p_2 g_{2x_1}], \qquad p_1(T) = 0, \tag{3.74}$$

$$\dot{p}_2 = -\frac{\partial H}{\partial x_2} = -[F_{x_2} + p_1 g_{1x_2} + p_2 g_{2x_2}], \qquad p_2(T) = 0. \tag{3.75}$$

The two-point boundary value equations and associated boundary conditions are then given by equations (3.72)–(3.75).

Newton–Raphson Equations for a Second-Order System. Using the Newton–Raphson method (References 1, 19), the unknown initial conditions are assumed to be given by

$$p_1(0) = c_1, \qquad p_2(0) = c_2. \tag{3.76}$$

Then expanding the boundary conditions

$$p_1(T, c_1, c_2) = 0, \qquad p_2(T, c_1, c_2) = 0 \tag{3.77}$$

in a Taylor series around the kth approximation and retaining only the linear terms

$$p_1(T, c_1^k, c_2^k) + (c_1^{k+1} - c_1^k) p_{1c_1} + (c_2^{k+1} - c_2^k) p_{1c_2} = 0, \tag{3.78}$$

$$p_2(T, c_1^k, c_2^k) + (c_1^{k+1} - c_1^k) p_{2c_1} + (c_2^{k+1} - c_2^k) p_{2c_2} = 0, \tag{3.79}$$

$$p_{ic_j} = \frac{\partial p_i}{\partial c_j}. \tag{3.80}$$

Solving for c_1^{k+1} and c_2^{k+1}, we obtain

$$c_1^{k+1} = c_1^k - (p_1 p_{2c_2} - p_2 p_{1c_2})/\Delta, \tag{3.81}$$

$$c_2^{k+1} = c_2^k - (p_2 p_{2c_1} - p_1 p_{2c_1})/\Delta, \tag{3.82}$$

where

$$\Delta = p_{1c_1} p_{2c_2} - p_{1c_2} p_{2c_1}. \tag{3.83}$$

The equations necessary to solve equations (3.81) and (3.82) are obtained by differentiating equations (3.72)–(3.75) with respect to c_1 and c_2. Differentiating with respect to c_1 we obtain

$$\dot{x}_{1c_1} = \sum_{i=1}^{N} g_{1x_i} x_{ic_1}, \tag{3.84}$$

$$\dot{x}_{2c_1} = \sum_{i=1}^{N} g_{2x_i} x_{ic_1}, \tag{3.85}$$

$$\dot{p}_{1c_1} = -\sum_{i=1}^{n} p_{ic_1} g_{ix_1} - \sum_{i=1}^{N} [F_{x_1x_i} + p_1 g_{1x_1x_i} + p_2 g_{2x_1x_i}] x_{ic_1}, \tag{3.86}$$

$$\dot{p}_{2c_1} = -\sum_{i=1}^{n} p_{ic_1} g_{ix_2} - \sum_{i=1}^{N} [F_{x_2x_i} + p_1 g_{1x_2x_i} + p_2 g_{2x_2x_i}] x_{ic_1}, \tag{3.87}$$

where for simplicity in expressing the summations, we define x_1, x_2, p_1, and p_2 equal to x_1, x_2, x_3, and x_4, respectively, with $N = 4$ and $n = 2$. The equations for the derivatives with respect to c_2 are the same as above with c_1 replaced by c_2. The initial conditions of these equations are all equal to zero except

$$p_{1c_1}(0) = 1, \qquad p_{2c_2}(0) = 1. \tag{3.88}$$

To obtain a numerical solution, an initial guess is made for initial conditions (3.76), i.e., usually c_1 and c_2 are set equal to zero. The 12 initial value differential equations (3.72)–(3.75), (3.84)–(3.87), and the equations differentiated with respect to c_2, are integrated from time $t = 0$ to $t = T$. A new set of values of c_1 and c_2 are calculated from equations (3.81) and (3.82) and the above sequence is repeated to obtain the second approximation, etc.

Automatic Derivative Evaluation. The partial derivatives of F, g_1, and g_2 in equations (3.74), (3.75), and (3.84)–(3.87) are computed automatically (Reference 15), so that once coded, the same equations for the partial derivatives apply for any F, g_1, and g_2. Equations (3.86) and (3.87) show that partial derivatives up to the second must be evaluated, i.e., $F_{x_1x_1}$, $g_{1x_1x_1}$, $g_{2x_1x_1}$, $F_{x_2x_1}$, etc. These elementary partial derivatives are used to form the partial derivatives such as $\dot{p}_{1c_1}$, and $\dot{p}_{2c_1}$, in equations (3.86) and (3.87), which are then integrated to obtain p_{1c_1} and p_{2c_1}.

Expressing F as a function of its arguments,

$$F = F(t, x_1, x_2, p_1, p_2), \tag{3.89}$$

Table 3.4. Number of Components of the Vectors, **F(I)**, **G1(I)**, or **G2(I)** for a Second-Order System, Newton–Raphson Method and Pontryagin's Maximum Principle

Function	Number of components
$F(t, \mathbf{x}, \mathbf{p})$, $g_1(t, \mathbf{x}, \mathbf{p})$ or $g_2(t, \mathbf{x}, \mathbf{p})$	1
First derivatives	5
Second derivatives	15

the partial derivatives are

$$F_1 = \frac{\partial F}{\partial t}, \qquad F_2 = \frac{\partial F}{\partial x_1}, \qquad F_3 = \frac{\partial F}{\partial x_2}, \qquad \text{etc.} \tag{3.90}$$

and

$$F_{11} = \frac{\partial^2 F}{\partial t\, \partial t}, \qquad F_{12} = \frac{\partial^2 F}{\partial t\, \partial x_1}, \qquad \text{etc.} \tag{3.91}$$

Not all the second derivatives need be calculated since the order of differentiation is immaterial, i.e., $F_{21} = F_{12}$, $F_{31} = F_{13}$, etc.

The user inputs the functions F, g_1, and g_2 into the program. The partial derivatives are then evaluated automatically. The functions F, g_1, and g_2 and all their derivatives are represented in the automatic solution program by the vectors **F(I)**, **G1(I)**, and **G2(I)**. These vectors consist of 21 components as shown in Table 3.4.

The definitions of the vector components are given in Table 3.5. Note that the order of the combinations of the second derivatives is such that in forming the derivatives, F_{ij}, first i is set equal to 1 while j is incremented from 1 to 5. Once j reaches its maximum limit of 5 then i is incremented to 2 while j is incremented from i to 5, etc. The last three second derivatives are not utilized so that an 18-component vector is actually sufficient.

Subroutine Linear. In subroutine LIN the variables, t, x_1, x_2, p_1, and p_2 are represented by $X(1)$, $X(2)$, $X(3)$, $X(4)$, and $X(5)$, respectively, and the corresponding vectors are represented by **A1**, **A2**, **A3**, **A4**, and **A5** as shown in Table 3.6. The first derivatives of the variables are equal to unity, while all other derivatives are equal to zero. From Table 3.5,

$$\mathbf{A1}(1) = t, \quad \mathbf{A2}(1) = x_1, \quad \mathbf{A3}(1) = x_2, \quad \mathbf{A4}(1) = p_1, \quad \mathbf{A5}(1) = p_2, \tag{3.92}$$

$$\mathbf{A1}(2) = 1, \quad \mathbf{A2}(3) = 1, \quad \mathbf{A3}(4) = 1, \quad \mathbf{A4}(5) = 1, \quad \mathbf{A5}(6) = 1 \tag{3.93}$$

and all other components of **A1**, . . . , **A5** are equal to zero.

Table 3.5. Definitions of the Vector Components for a Second-Order System, Newton–Raphson Method and Pontryagin's Maximum Principle

Vector component L	Derivative arguments		Symbol[a]
	i	j	
1	—		z
2	1		z_t
3	2		z_{x_1}
4	3		z_{x_2}
5	4		z_{p_1}
6	5		z_{p_2}
7	1	1	z_{tt}
8	1	2	z_{tx_1}
9	1	3	z_{tx_2}
10	1	4	z_{tp_1}
11	1	5	z_{tp_2}
12	2	2	$z_{x_1x_1}$
13	2	3	$z_{x_1x_2}$
14	2	4	$z_{x_1p_1}$
15	2	5	$z_{x_1p_2}$
16	3	3	$z_{x_2x_2}$
17	3	4	$z_{x_2p_1}$
18	3	5	$z_{x_2p_2}$
19	4	4	$z_{p_2p_1}$
20	4	5	$z_{p_1p_2}$
21	5	5	$z_{p_2p_2}$

[a] $z_i = \partial z/\partial i$, $z_{ij} = \partial^2 z/\partial i \partial j$, where z is a scalar variable equal to t, x_1, x_2, p_1, or p_2 or a function of the variables, such as the integrand F. Only the first 18 components are used in the program in Section 3.5.2.

Table 3.6. FORTRAN Representation of the Variables for a Second-Order System, Newton–Raphson Method and Pontryagin's Maximum Principle

Variables	FORTRAN program	
	Variables	Vectors
t	$X(1)$	A1
x_1	$X(2)$	A2
x_2	$X(3)$	A3
p_1	$X(4)$	A4
p_2	$X(5)$	A5

3.2.2. Gradient Method

The gradient method for obtaining the numerical solution of optimal control problems with Pontryagin's maximum principle is described in this section. The optimal control is obtained with either (i) the evaluation of the minimum cost functional or (ii) the Newton–Raphson recurrence relation. The latter is used to obtain the solution to the general nth-order nonlinear system optimal control problem.

3.2.2.1. *Gradient Method with Evaluation of Minimum Cost Functional*

Consider the scalar optimal control problem of minimizing the cost functional

$$J = \int_0^T F(t, x, y)\,dt \tag{3.94}$$

subject to the nonlinear differential constraint and initial condition

$$\dot{x} = f(t, x, y), \qquad x(0) = x_0, \tag{3.95}$$

where x is the scalar state variable, y is the scalar control, and f is a nonlinear function of x and y. Utilizing Pontryagin's maximum principle, the Hamiltonian function is

$$H = F(t, x, y) + p(t)f(t, x, y), \tag{3.96}$$

where the costate variable, $p(t)$, is the solution of the differential equation and boundary condition

$$\dot{p} = -\frac{\partial H}{\partial x}, \qquad p(T) = 0. \tag{3.97}$$

The control, $y(t)$, that minimizes H must satisfy the equation

$$\frac{\partial H}{\partial y} = 0. \tag{3.98}$$

Using the gradient method along with the table method, the numerical solution of the above optimal control problem can be obtained automatically. The user of the program need only input the integrand, F, of the cost functional, the Hamiltonian, H, and specify the initial conditions and the terminal time, T.

Gradient Method for a First-Order System. From equations (3.95)–(3.98), the equations to be solved for the optimal control problem are

$$\dot{x} = H_p(t, x, p, y) = f(t, x, y), \qquad x(0) = x_0, \tag{3.99}$$

$$\dot{p} = -H_x(t, x, p, y), \qquad p(T) = 0, \tag{3.100}$$

$$H_y(t, x, p, y) = 0, \tag{3.101}$$

where the subscripts are used to define the partial derivatives, i.e.,

$$H_p = \frac{\partial H(t, x, p, y)}{\partial p}, \text{ etc.} \tag{3.102}$$

Assume now that $y(t)$ is an initial guess of the solution for the optimal control. Then it can be shown that the incremental first-order change in the cost functional caused by a change in $y(t)$ of $\Delta y(t)$ is given by (Reference 4)

$$\Delta J = \int_0^T H_y(t, x, p, y)\Delta y\, dt. \tag{3.103}$$

To obtain the largest change in ΔJ, let

$$\Delta y(t) = -K[H_y(t, x, p, y)], \tag{3.104}$$

where K is a nonnegative constant or function of time.

Equations (3.103) and (3.104) can be verified as follows (References 3 and 4). Adjoin equation (3.95) to equation (3.94) and substitute the Hamiltonian function given by equation (3.96). Then

$$J = \int_0^T \{F(t, x, y) + p(t)[f(t, x, y) - \dot{x}]\}\, dt$$

$$= \int_0^T (H - p\dot{x})\, dt. \tag{3.105}$$

Integrating the second term on the right-hand side of equation (3.105) by parts, we obtain

$$J = p(0)x(0) - p(T)x(T) + \int_0^T (H + \dot{p}x)\, dt. \tag{3.106}$$

The variation in J due to the variation in y is then

$$\Delta J = p(0)\Delta x(0) - p(T)\Delta x(T) + \int_0^T [(H_x + \dot{p})\Delta x + H_y\Delta_y]\, dt. \tag{3.107}$$

This becomes equation (3.103) since

$$\dot{p} = -H_x, \qquad p(T) = 0 \tag{3.108}$$

and the initial condition is fixed, i.e.,

$$\Delta x(0) = 0. \tag{3.109}$$

In order to verify equation (3.104), approximate the integral given by equation (3.103) by the summation

$$\Delta J = \sum_{i=1}^{N} H_y{}^i \Delta y^i \Delta t^i, \qquad N = \frac{T}{\Delta t}. \tag{3.110}$$

Then to obtain the largest decrease in the cost functional J, the summation ΔJ should be negative. Therefore select

$$\Delta y^i = -KH_y{}^i \tag{3.111}$$

which yields the negative quantity

$$\Delta J = \sum_{i=1}^{N} - K(H_y{}^i)^2 \Delta t^i. \tag{3.112}$$

The solution to the optimal control problem is then obtained as follows. Assume the control is $y^N(t)$. Integrate $\dot{x}$ forward from $t = 0$ to $t = T$. Note that $\dot{x}(t)$ in equation (3.99) is not dependent directly on $p(t)$. Once $x(t)$ is obtained, integrate $\dot{p}$ backward from $t = T$ to $t = 0$, computing $H_y(t, x, p, y)$ along the way. The new value of $y(t)$ for the next successive approximation of the solution is then given by

$$y^{N+1}(t) = y^N(t) + \Delta y^N(t), \tag{3.113}$$

where $\Delta y^N(t)$ is obtained from equation (3.104).

Reference 4 suggests that in order to select K, let

$$\Delta y_i{}^N(t) = -K_i H_y(t, x^N, p^N, y^N), \qquad i = 1, \ldots, 4, \tag{3.114}$$

where $K_i = \frac{1}{2}$, 1, 2, and 4 times the previous value of K. The resulting four values of $y_i{}^{N+1}(t)$,

$$y_i{}^{N+1} = y^N(t) + \Delta y_i{}^N(t) \tag{3.115}$$

are used to integrate $\dot{x}(t)$ and to evaluate the cost functional

$$J_i^{N+1} = \int_0^T F(t, x_i^{N+1}, y_i^{N+1})\, dt. \tag{3.116}$$

The value of K_i which produces the smallest J_i^{N+1} is then used for the $N + 1$ approximation.

Automatic Derivative Evaluation. To use the automatic method of solution, the Hamiltonian H must be input into the program as described in Section 3.4. The partial derivatives, H_t, H_x, H_p, and H_y are calculated automatically. Equation (3.101), i.e., $H_y = 0$, does not have to be solved explicitly for the control $y(t)$. The control $y^N(t)$ is computed successively using the gradient.

For this method, only the first derivatives are required (Reference 16). The Hamiltonian, $H(t, x, p, y)$, and all of its derivatives are represented in the automatic solution program by the vector **HA(I)**. The number of components of this vector is 5 as shown in Table 3.7.

Equation (3.116) is integrated using Simpson's rule. Thus in addition to the Hamiltonian, H, the integrand F must be formed. The functions H and F are input into the program in the INPUT subroutine.

To obtain the numerical results a fourth-order Runge–Kutta integration method is used typically with grid intervals $\Delta t = 1/100$. The values of $x(t)$ are stored during the forward integration and $H_y(t, x, p, y)$ is computed and stored during the backward integration. The control $y(t)$ is then computed and stored using equation (3.115) with the appropriate value of K. The procedure is then repeated with the forward integration of $\dot{x}$, etc.

The stored values of $y(t)$ are used during the forward integration with linear interpolation between the grid intervals. Both the stored values of the state, $x(t)$, and the control, $y(t)$, are used during the backward integration with linear interpolation between the grid intervals.

Table 3.7. Number of Components of the Vector **HA(I)** or **F(I)** for a First-Order System, Gradient Method with Evaluation of Minimum Cost Functional

Function	Number of components
$H(t, x, p, y)$ or $F(t, x, y)$	1
First derivatives	4

Table 3.8. Definitions of the Vector Components for a First-Order System, Gradient Method with Evaluation of Minimum Cost Functional

Vector component L	Derivative argument i	Symbol[a]	Example $z = A1(L)$
1	—	z	$A1(1) = t$
2	1	z_t	$A1(2) = 1$
3	2	z_x	$A1(3) = 0$
4	3	z_p	$A1(4) = 0$
5	4	z_y	$A1(5) = 0$

[a] $z_i = \partial z/\partial i$, where z is a scalar variable equal to t, x, p, or y or a function of the variables, such as the Hamiltonian H.

The definitions of the vector components are given in Table 3.8.

Subroutine Linear. In subroutine LIN the variables, t, x, p, and y are represented by $X(1)$, $X(2)$, $X(3)$, and $X(4)$, respectively, and the corresponding vectors are represented by **A1**, **A2**, **A3**, and **A4** as shown in Table 3.9. The first derivatives of the variables are equal to unity, while all other derivatives are equal to zero. From Table 3.8 we have

$$\mathbf{A1}(1) = t, \qquad \mathbf{A2}(1) = x, \qquad \mathbf{A3}(1) = p, \qquad \mathbf{A4}(1) = y, \tag{3.117}$$

$$\mathbf{A1}(2) = 1, \qquad \mathbf{A2}(3) = 1, \qquad \mathbf{A3}(4) = 1, \qquad \mathbf{A4}(5) = 1 \tag{3.118}$$

and all other components of **A1**, . . . , **A4** are equal to zero.

Table 3.9. FORTRAN Representation of the Variables for a First-Order System, Gradient Method with Evaluation of Minimum Cost Functional

Variables	FORTRAN program	
	Variables	Vectors
t	$X(1)$	**A1**
x	$X(2)$	**A2**
p	$X(3)$	**A3**
y	$X(4)$	**A4**

3.2.2.2. Gradient Method with Newton–Raphson Recurrence Relation

Consider the nth-order system optimal control problem of extremizing the cost functional

$$J = \int_0^T F(t, \mathbf{x}, y)\, dt \tag{3.119}$$

subject to the nonlinear differential constraints and initial conditions

$$\dot{\mathbf{x}} = \mathbf{f}(t, \mathbf{x}, y), \qquad \mathbf{x}(0) = \mathbf{x}_0, \tag{3.120}$$

where $\mathbf{x}$ is an n-dimensional state vector, y is the scalar control, and $\mathbf{f}$ is an n-dimensional vector the components of which, $f_1, f_2, \ldots, f_n$, are nonlinear functions of $\mathbf{x}$ and y.

Utilizing Pontryagin's maximum principle, the Hamiltonian function is

$$H = F(t, \mathbf{x}, y) + \mathbf{p}^T(t)\mathbf{f}(t, \mathbf{x}, y) \tag{3.121}$$

where the costate variables are the solutions of the differential equations and boundary conditions

$$\mathbf{p} = -\frac{\partial H}{\partial \mathbf{x}}, \qquad \mathbf{p}(T) = 0. \tag{3.122}$$

The control, $y(t)$, that minimizes H must satisfy the equation

$$\frac{\partial H}{\partial y} = 0. \tag{3.123}$$

Gradient Method for an Nth-Order System. From equations (3.120)–(3.123), the equations to be solved for the optimal control problem are

$$\dot{\mathbf{x}} = H_{\mathbf{p}}(t, \mathbf{x}, \mathbf{p}, y) = \mathbf{f}(t, \mathbf{x}, y), \qquad \mathbf{x}(0) = \mathbf{x}_0, \tag{3.124}$$

$$\dot{\mathbf{p}} = -H_{\mathbf{x}}(t, \mathbf{x}, \mathbf{p}, y), \qquad p(T) = \mathbf{0}, \tag{3.125}$$

$$H_y(t, \mathbf{x}, \mathbf{p}, y) = 0, \tag{3.126}$$

where $\mathbf{x}$, $\mathbf{p}$, and $\mathbf{f}$ are n-dimensional vectors and the subscripts are used to define the partial derivatives, i.e.,

$$H_{\mathbf{p}} = \frac{\partial H(t, \mathbf{x}, \mathbf{p}, y)}{\partial \mathbf{p}}, \qquad \text{etc.} \tag{3.127}$$

Assume now that $y(t)$ is an initial guess of the solution for the optimal control. Then as was shown for the scalar case in Section 3.2.2.1, the incremental first-order change in the cost functional caused by a change in $y(t)$ of $\Delta y(t)$ is given by (Reference 4)

$$\Delta J = \int_0^T H_y(t, \mathbf{x}, \mathbf{p}, y)\Delta y \, dt. \tag{3.128}$$

To obtain the largest change in ΔJ, let

$$\Delta y(t) = -K[H_y(t, \mathbf{x}, \mathbf{p}, y)], \tag{3.129}$$

where K is a nonnegative constant or function of time.

The solution is then obtained as follows. Assume the control is $y^N(t)$. Integrate $\dot{\mathbf{x}}$ forward from $t = 0$ to $t = T$. Note that $\dot{\mathbf{x}}(t)$ in equation (3.124) is not dependent directly on $\mathbf{p}(t)$. Once $\mathbf{x}(t)$ is obtained, integrate $\dot{\mathbf{p}}$ backward from $t = T$ to $t = 0$, computing $H_y(t, \mathbf{x}, \mathbf{p}, y)$ along the way. The new value of $y(t)$ for the next successive approximation of the solution is then given by

$$y^{N+1}(t) = y^N(t) + \Delta y^N(t), \tag{3.130}$$

where $\Delta y^N(t)$ is obtained from equation (3.129).

In order to select the value of K, four values of K were chosen in Section 3.2.2.1 such that $K_i = \frac{1}{2}$, 1, 2, and 4 times the previous value of K. The resulting four values of $y_i^{N+1}(t)$ were then used to integrate $\dot{\mathbf{x}}_i(t)$ and to evaluate the cost functional J_i^{N+1}, where

$$J_i^{N+1} = \int_0^T F(t, \mathbf{x}_i^{N+1}, y_i^{N+1}) \, dt \tag{3.131}$$

and

$$y_i^{N+1}(t) = y^N(t) - K_i H_y(t, \mathbf{x}^N, \mathbf{p}^N, y^N), \qquad i = 1, \ldots, 4. \tag{3.132}$$

The value of K_i which produces the smallest J_i^{N+1} is used for the $N + 1$ approximation.

In this section, the Newton–Raphson recurrence relation is used to obtain the optimal control. Expanding equation (3.126) in a Taylor series around the nth approximation and retaining only the linear terms, we obtain

$$H_y + (y^{N+1} - y^N)H_{yy} = 0. \tag{3.133}$$

Solving for $y^{N+1}(t)$ we obtain the Newton–Raphson recurrence relation (References 17 and 18):

$$y^{N+1}(t) = y^N(t) - \left(\frac{1}{H_{yy}}\right)H_y, \tag{3.134}$$

where $H_{yy}(t, \mathbf{x}, \mathbf{p}, y)$ is evaluated at each integration step, Δt. This method is more accurate than using a constant value of K for the whole trajectory and results in a considerable savings in computation time since $\dot{\mathbf{x}}(t)$ and the integrand of the cost functional, $F(t, \mathbf{x}, y)$, do not have to be integrated for the four different values of K at each iteration. The evaluations of H_y and H_{yy} are obtained automatically using the table method.

Automatic Derivative Evaluation. To use the automatic method of solution, the Hamiltonian H must be input into the program as described in Section 3.4. The partial derivatives, H_t, $H_{\mathbf{x}}$, $H_{\mathbf{p}}$, H_y, and H_{yy} are calculated automatically. Equation (3.126), i.e., $H_y = 0$, does not have to be solved explicitly for the control $y(t)$. The control $y^N(t)$ is computed successively using the Newton–Raphson recurrence relation.

To obtain the numerical results a fourth-order Runge–Kutta integration method is used typically with grid intervals $\Delta t = 1/100$. The values of $\mathbf{x}(t)$ are stored during the forward integration and H_y/H_{yy} is computed and stored during the backward integration. The control $y(t)$ is then computed and stored using equation (3.134). The procedure is then repeated with the forward integration of $\dot{\mathbf{x}}$, etc.

The stored values of $y(t)$ are used during the forward integration with linear interpolation between the grid intervals. Both the stored values of the state, $\mathbf{x}(t)$, and the control, $y(t)$, are used during the backward integration with linear interpolation between the grid intervals.

The partial derivatives of the Hamiltonian, H, in equations (3.124)–(3.126) are computed automatically, so that once coded, the same equations for the partial derivatives apply for any H. For example, expressing H as a function of its arguments,

$$H = H(t, x_1, x_2, \ldots, x_n, p_1, p_2, \ldots, p_n, y), \tag{3.135}$$

we obtain the partial derivatives

$$H_1 = \frac{\partial H}{\partial t}, \qquad H_2 = \frac{\partial H}{\partial x_1}, \qquad \text{etc.} \tag{3.136}$$

The vector **HA(I)** in the table method consists of $2n + 4$ components as shown in Table 3.10. The first component corresponds to $H(t, \mathbf{x}, \mathbf{p}, y)$, the next $2n + 2$ components correspond to the first partial derivatives, and the last component corresponds to H_{yy}. The definitions of the vector components are given in Table 3.11.

Table 3.10. Number of Components of the Vector, **HA(I)** for an *n*th-Order System, Gradient Method with Newton–Raphson Recurrence Relation

Function	Number of components
$H(t, \mathbf{x}, \mathbf{p}, y)$	1
First derivatives	$2n + 2$
Second derivatives	1

Subroutine Linear. In subroutine LIN the variables, t, x_1, x_2, ..., x_n, p_1, p_2, ..., p_n, and y are represented by $X(1)$, $X(2)$, ..., $X(2n + 2)$, respectively, and the corresponding variables and their derivatives are represented by the rows of the $(2n + 2) \times (2n + 4)$ matrix A. The vectors corresponding to the rows of A are represented in subroutine INPUT by **T**, **X**1, **X**2, ..., **X***n*, **P**1, **P**2, ..., **P***n*, and **Y** as shown in Table 3.12. The

Table 3.11. Definitions of the Vector Components for an *n*th-Order System, Gradient Method with Newton–Raphson Recurrence Relation

Vector component L	Derivative argument i	Symbol[a]	Example $z = X1(L)$
1	—	z	$X1(1) = x_1$
2	1	z_t	$X1(2) = 0$
3	2	z_{x_1}	$X1(3) = 1$
4	3	z_{x_2}	$X1(4) = 0$
⋮			
$n + 2$	$n + 1$	z_{x_n}	$X1(n + 2) = 0$
$n + 3$	$n + 2$	z_{p_1}	$X1(n + 3) = 0$
$n + 4$	$n + 3$	z_{p_2}	$X1(n + 4) = 0$
⋮			
$2n + 2$	$2n + 1$	z_{p_n}	$X1(2n + 2) = 0$
$2n + 3$	$2n + 2$	z_y	$X1(2n + 3) = 0$
$2n + 4$	$2n + 3$	z_{yy}	$X1(2n + 4) = 0$

[a] $z_i = \partial z/\partial i$, where z is a scalar variable equal to t, x_1, x_2, ..., x_n, p_1, p_2, ..., p_n, or y or a function of the variables, such as the Hamiltonian H.

Table 3.12. FORTRAN Representation of the Variables for an *n*th-Order System, Gradient Method with Newton–Raphson Recurrence Relation

Variables	FORTRAN program	
	Variables	Vectors
t	$X(1)$	**T**
x_1	$X(2)$	**X1**
$\vdots$		
x_n	$X(n+1)$	**X***n*
p_n	$X(n+2)$	**P1**
$\vdots$		
p_n	$X(2n+1)$	**P***n*
y	$X(2n+2)$	**Y**

first derivatives of the variables are equal to unity, while all the other derivatives are equal to zero. From Table 3.11,

$$\mathbf{T}(1) = t,\ \mathbf{X}1(1) = x_1,\ \ldots,\ \mathbf{X}n(1) = x_n,\ \mathbf{P}1(1) = p_1,\ \ldots,\ \mathbf{P}n(1) - p_n,\ \mathbf{Y}(1) = y, \tag{3.137}$$

$$\mathbf{T}(2) = 1,\ \mathbf{X}1(3) = 1,\ \ldots,\ \mathbf{X}n(n+2) = 1,\ \mathbf{P}1(n+3) = 1,\ \ldots,\ \mathbf{P}n(2n+2) = 1,\ \mathbf{Y}(2n+3) = 1 \tag{3.138}$$

and all other components of the vectors are equal to zero.

3.3. Description of the Subroutines

In order to compute the derivatives automatically (References 13–18), an L-component vector must be computed for each of the m variables, t, $x_1, x_2, \ldots, x_n, p_1, p_2, \ldots, p_n$, and y and for each of the functions of the m variables, such as the sums, the product, the square root, the quotient, etc. The subroutines required to compute these vectors are as follows: (1) linear, (2) constant, (3) add, (4) multiplication, (5) division, and (6) function. The linear subroutine was described in Section 3.2. Subroutines (2)–(6) are described below.

The derivatives to be evaluated are

$$G_i = \frac{\partial G}{\partial i}, \qquad G_{ij} = \frac{\partial^2 G}{\partial i\, \partial j}, \qquad G_{ijk} = \frac{\partial^3 G}{\partial i\, \partial j\, \partial k} \tag{3.139}$$

where G is a function of the variables and $i, j, k = 1, 2, \ldots, m$ corresponding to the m variables $t, x_1, x_2, \ldots, x_n, p_1, p_2, \ldots, p_n, y$. The number of variables and the order of the derivatives required depend on the system order of the problem and the numerical method of solution used. The variables are defined in Tables 3.3, 3.6, 3.9, and 3.12, and the corresponding derivatives are defined in Tables 3.2, 3.5, 3.8, and 3.11. For example for the first-order system using the Newton–Raphson method with the Euler–Lagrange equations only the variables t, x, and $\dot{x}$ are required. However, derivatives up to the third must be evaluated as shown in Tables 3.2 and 3.3.

3.3.1. Subroutine Constant

In subroutine CONST, a constant term, such as a constant coefficient multiplier in the system equation, is represented by CON and the corresponding vector is represented by D. The component $D(1) = \text{CON}$ and all of the derivatives are equal to zero.

3.3.2. Subroutine Add

In subroutine ADD, two functions, D and E, and their derivatives are added:

$$G = D + E, \qquad \text{functions,} \tag{3.140}$$

$$G_i = D_i + E_i, \qquad \text{first derivatives,} \tag{3.141}$$

$$G_{ij} = D_{ij} + E_{ij}, \qquad \text{second derivatives,} \tag{3.142}$$

$$G_{ijk} = D_{ijk} + E_{ijk}, \qquad \text{third derivatives.} \tag{3.143}$$

Using the table method the two vectors, $D(I)$ and $E(I)$, are simply added:

$$G(I) = D(I) + E(I), \qquad I = 1, \ldots, L. \tag{3.144}$$

Subtraction is obtained by multiplying E by -1 and then adding.

3.3.3. Subroutine Multiplication

In subroutine MULT, two functions, A and B are multiplied:

$$E = AB. \tag{3.145}$$

The first, second, and third derivatives are

$$E_i = A_i B + A B_i, \tag{3.146}$$

$$E_{ij} = A_{ij} B + A_i B_j + A_j B_i + A B_{ij}, \tag{3.147}$$

$$E_{ijk} = A_{ijk} B + A_{ij} B_k + A_{ik} B_j + A_i B_{jk} + A_{jk} B_i + A_j B_{ik} + A_k B_{ij} + A B_{ijk}. \tag{3.148}$$

Using the table method, the first derivatives are formed as follows:

$$\begin{gathered} E(I+1) = A(I+1)B(1) + A(1)B(I+1), \\ I = 1, \ldots, m, \qquad m = \text{number of variables}. \end{gathered} \tag{3.149}$$

The complete set of second- and third-order derivatives is required only for the Newton–Raphson method using the Euler–Lagrange equations as shown by the definitions of the vector components in Table 3.2.

Using the table method, the second derivatives are formed as follows:

$$\begin{gathered} E(L) = A(L)B(1) + A(I+1)B(J+1) + A(J+1)B(I+1) + A(1)B(L), \\ I = 1, \ldots, m, \qquad J = I, \ldots, m. \end{gathered} \tag{3.150}$$

The component, L, is incremented by one each time I or J is incremented as shown in Table 3.2.

The third derivatives are obtained as follows:

$$\begin{aligned} E(L) = {} & A(L)B(1) + A(IJ)B(K+1) + A(IK)B(J+1) + A(I+1)B(JK) \\ & + A(JK)B(I+1) + A(J+1)B(IK) + A(K+1)B(IJ) + A(1)B(L), \\ & I = 1, \ldots, m, \qquad J = I, \ldots, m, \qquad K = J, \ldots, m. \end{aligned} \tag{3.151}$$

The component, L, is incremented by one each time I, J, or K is incremented. The component IJ is incremented by one each time I or J is incremented. The component IK is incremented by one each time I or K is incremented. To detect when the latter occurs, a DO loop for $I1 = 1, \ldots, m$; $K1 = I1, \ldots, m$ is required. The component JK is incremented by one each time J or K is incremented. To detect when J or K is incremented, a DO loop for $J1 = 1, \ldots, m$; $K1 = J1, \ldots, m$ is required.

Up to only the second derivatives are required for the Newton–Raphson method using Pontryagin's maximum principle as shown by the definitions of the vector components in Table 3.5. Using the gradient method with the evaluation of the minimum cost functional, only the first derivatives need be evaluated, as shown in Table 3.8. The gradient method with the Newton–Raphson recurrence relation requires only the first derivatives and the single second derivative

$$E_{yy} = A_{yy}B + 2A_yB_y + AB_{yy} \tag{3.152}$$

to be evaluated as a special case since none of the other second derivatives are required for the solution. This is shown by the definitions of the vector components in Table 3.11.

3.3.4. Subroutine Division

In subroutine DIV, the derivatives of the quotient, B/U, are obtained. In this subroutine, the function B is the numerator and the function U is the denominator. The derivatives of the reciprocal, $R = 1/U$, are calculated and then subroutine MULT is called to obtain the function $F1 = B*R$ and its derivatives.

Consider the function

$$R = f(u). \tag{3.153}$$

Then the first, second, and third derivatives are

$$R_i = f'(u)u_i, \tag{3.154}$$

$$R_{ij} = f''(u)u_ju_i + f'(u)u_{ij}, \tag{3.155}$$

$$R_{ijk} = f'''(u)u_ku_ju_i + f''(u)u_{jk}u_i + f''(u)u_ju_{ik} + f''(u)u_ku_{ij} + f'(u)u_{ijk}, \tag{3.156}$$

where

$$f(u) = u^{-1}, \tag{3.157}$$

$$f'(u) = -u^{-2}, \tag{3.158}$$

$$f''(u) = 2u^{-3}, \tag{3.159}$$

$$f'''(u) = -6u^{-4}. \tag{3.160}$$

Equations (3.157)–(3.160) are evaluated in subroutine DIV. Equations (3.153)–(3.156) are then evaluated by calling subroutine DER. The method of calculating the L-component vector of the function R and its derivatives in DER is similar to the method used in MULT.

3.3.5. Subroutine Function

In addition to the arithmetic operations, add, subtract, multiply, and divide described above, certain special functions such as the square root can be used as primitives in building up other functions. Subroutine DER can be used to obtain the derivatives of any function of a single variable. For example to obtain the square root, subroutine SR evaluates the equations

$$f(u) = u^{1/2} \tag{3.161}$$

$$f'(u) = \tfrac{1}{2} u^{-1/2} \tag{3.162}$$

$$f''(u) = -\tfrac{1}{4} u^{-3/2} \tag{3.163}$$

$$f'''(u) = \tfrac{3}{8} u^{\;5/2} \tag{3.164}$$

Subroutine DER is then called to evaluate equations (3.153)–(3.156), where $f(u)$ and its derivatives are defined as above.

To obtain the $\sin(u)$, subroutine SSIN evaluates the equations

$$f(u) = \sin(u) \tag{3.165}$$

$$f'(u) = \cos(u) \tag{3.166}$$

$$f''(u) = -\sin(u) \tag{3.167}$$

$$f'''(u) = \cos(u) \tag{3.168}$$

Subroutine DER is then called to evaluate equations (3.153)–(3.156).

3.4. Examples of Optimal Control Problems

Examples of optimal control problems are given in this section using the numerical methods of solution described in Section 3.2. The automatic method of solution is used with (i) the Newton–Raphson method or (ii) the gradient method.

The procedures for entering the optimal control problem examples into the automatic solution program via the INPUT subroutine are described in detail. Numerical results are given for first-, second-, and third-order nonlinear systems. Extensions to nth-order systems are discussed in this section and in Section 3.5.3.

3.4.1. Newton–Raphson Method

The Newton–Raphson method with the Euler–Lagrange equations is used to obtain the numerical solutions for the simplest problem in the calculus of variations and the first-order nonlinear system examples in Sections 3.4.1.1 and 3.4.1.2. The Newton–Raphson method with the equations of Pontryagin's maximum principle is used to obtain the numerical solutions for the examples of the second-order nonlinear systems and the first-order systems with integral constraints in Sections 3.4.1.3 and 3.4.1.4.

3.4.1.1. Simplest Problem in the Calculus of Variations

The Newton–Raphson method with the Euler–Lagrange equations, as described in Section 3.2.1.1, is used in this section to obtain the numerical solution for the simplest problem in the calculus of variations (Reference 13). The simplest problem in the calculus of variations involves the extremization of an integral

$$J = \int_a^b F(t, x, \dot{x})\, dt \tag{3.169}$$

with boundary conditions

$$x(a) = c, \qquad x(b) = d, \tag{3.170}$$

where $x = x(t)$, $a \leq t \leq b$, is an unknown function of the independent variable t. The integrand F is a given function of t, x, and $\dot{x}$, where

$$\dot{x} = dx/dt. \tag{3.171}$$

To obtain the extremization, the unknown function $x(t)$ must satisfy the Euler–Lagrange equation

$$F_x - \frac{d}{dt} F_{\dot{x}} = 0 \tag{3.172}$$

with boundary conditions

$$x(a) = c, \qquad x(b) = d, \tag{3.173}$$

where

$$F_x = \frac{\partial F}{\partial x}. \tag{3.174}$$

As an example, consider an application of Fermat's principle (Reference 19). Let the (t, x)-plane represent an optically inhomogeneous medium.

The problem is to find the path that a light particle takes in passing from point $t = a$ to point $t = b$. The integrand in equation (3.169) is given by

$$F(t, x, \dot{x}) = \frac{(1 + \dot{x}^2)^{1/2}}{x} \tag{3.175}$$

and the boundary conditions are

$$x(a) = c, \qquad x(b) = d. \tag{3.176}$$

Using the automatic method of solution, assume that the boundary conditions are

$$x(1) = 1, \qquad x(2) = 2 \tag{3.177}$$

and that the initial approximation of the slope is

$$\dot{x}(1) = 0.1. \tag{3.178}$$

The only inputs required from the user are the integrand and the boundary conditions. The boundary conditions are specified in the main program. The integrand is input in subroutine INPUT. The FORTRAN listing of the INPUT subroutine is given in Table 3.13. The integrand is represented by the vector **F1**. The other variables and vectors are as defined in

Table 3.13. FORTRAN Listing of the INPUT Subroutine for the Simplest Problem in the Calculus of Variations, Newton–Raphson Method

Listing	Purpose
Subroutine INPUT(X1,X2,X3, CON,A,B,C,F1)	
Dimension A(20),B(20),C(20), D(20),E(20)	
Dimension G(20),S(20),F1(20)	
Call LIN(X1,X2,X3,A,B,C)	Defines the vectors corresponding to t, x, and $\dot{x}$.
Call CONST(CON,D)	Defines the vector corresponding to constant = 1.
Call MULT(C,C,E)	Multiplies $\dot{x}$ times $\dot{x}$ to form $E = \dot{x}^2$.
Call ADD(D,E,G)	Forms the sum $G = (1 + \dot{x}^2)$.
Call SR(G,S)	Takes the square root of G to form $S = (1+\dot{x}^2)^{1/2}$.
Call DIV(B,S,F1)	Divides S by x to form $F1 = (1 + \dot{x}^2)^{1/2}/x$.
Return	
End	

Table 3.14. Numerical Results after Six Iterations for the Simplest Problem in the Calculus of Variations, Newton–Raphson Method

t	x	$\dot{x}$
1.000000	1.000000	2.000000
1.100000	1.178983	1.611559
1.200000	1.326650	1.356801
1.300000	1.452584	1.170328
1.400000	1.562050	1.024295
1.500000	1.658312	0.9045340
1.600000	1.743560	0.8029551
1.700000	1.819341	0.7145447
1.800000	1.886796	0.6359988
1.900000	1.946792	0.5650321
2.000000	2.000000	0.5000000

Table 3.3. The vector, **D**, represents the constant = 1. The vectors **E**, **G**, and **S** represent intermediate variables. The vectors consist of the 20 components given in Table 3.2. The complete program listing is given in Section 3.5.1.

The differential equations are integrated using a fourth-order Runge–Kutta integration method with grid intervals $\Delta t = 1/100$. The equations to be integrated, equations (3.56)–(3.59), are represented in the integration subroutine by the four-component vectors **Z** and **DZ** for x_i and its derivatives respectively.

Table 3.14 shows the numerical results after six iterations. These results can be compared with the exact solution. It is known that the optimizing curves are arcs of circles with their centers on the t axis. For the given example, the equation of the solution is

$$(t - 3)^2 + x^2 = 5, \tag{3.179}$$

from which it can be shown that the numerical results are in agreement with the exact solution to seven digits.

3.4.1.2. *First-Order Nonlinear Systems*

In this section, an example of a first-order system optimal control problem is considered (Reference 14). As in the previous example and described in Section 3.2.1.1, the Newton–Raphson method with the Euler–

Lagrange equations is used to obtain the numerical solution. Consider the cost functional

$$J = \int_0^T (x^2 + y^2)\, dt \tag{3.180}$$

subject to the nonlinear differential constraint and initial condition

$$\dot{x} = -ax - bx^2 + y, \qquad x(0) = c. \tag{3.181}$$

Solving equation (3.181) for y and substituting into equation (3.180) yields the integrand

$$F(t, x, \dot{x}) = x^2 + (\dot{x} + ax + bx^2)^2. \tag{3.182}$$

Using the automatic method of solution assume that

$$a = 1, \qquad b = 0.1, \tag{3.183}$$

$$x(0) = 1, \qquad T = 1 \tag{3.184}$$

and that the initial approximation of the slope is

$$\dot{x}(0) = -1. \tag{3.185}$$

The only inputs required from the user are the integrand, the initial conditions and the terminal time, T. The initial conditions and the terminal time are specified in the main program. The integrand is input in subroutine INPUT. The FORTRAN listing of the INPUT subroutine is given in Table 3.15. The integrand is represented by the vector **F1**. The other variables and vectors are defined as in Table 3.3. The vectors **D** and **G** represent the coefficients a and b. The vectors **E1**, **E2**, ..., **E6** represent intermediate variables. The vectors consists of the 20 components given in Table 3.2.

Table 3.16 shows the numerical results after four iterations. The program computes x and $\dot{x}$ as a function of t. The optimal control is obtained from the equation

$$y = \dot{x} + ax + bx^2. \tag{3.186}$$

Considerably more effort is required from the user if the automatic method of solution is not used. As an exercise, it is informative to compare the above with the equations and the numerical results obtained by solving for the Euler–Lagrange and Newton–Raphson equations by hand, and substituting these equations in place of the INPUT subroutine and in the

Table 3.15. FORTRAN Listing of the INPUT Subroutine for the First-Order Nonlinear System, Newton–Raphson Method

Listing	Purpose
Subroutine INPUT(X1,X2,X3, CON,A,B,C,F1)	
Dimension A(20),B(20),C(20), D(20),G(20),F1(20)	
Dimension E1(20),E2(20), E3(20),E4(20),E5(20),E6(20)	
Call LIN(X1,X2,X3,A,B,C)	Defines the vectors corresponding to t, x, and $\dot{x}$.
Call CONST(CON,D)	Defines the vector corresponding to coefficient $a = 1$.
Call MULT(D,B,E1)	Multiplies a times x to form $E1 = ax$.
B1 = 0.1	Defines the coefficient $b = 0.1$.
Call MULT(B,B,E2)	Multiplies x times x to form $E2 = x^2$.
Call CONST(B1,G)	Defines the vector corresponding to coefficient $b = 0.1$.
Call MULT(G,E2,E3)	Multiplies b times x^2 to form $E3 = bx^2$.
Call ADD(E1,E3,E4)	Forms the sum $E4 = ax + bx^2$.
Call ADD(C,E4,E5)	Forms the sum $E5 = y = \dot{x} + ax + bx^2$.
Call MULT(E5,E5,E6)	Multiplies y times y to form $E6 = (\dot{x} + ax + bx^2)^2$.
Call ADD(E2,E6,F1)	Forms the sum $F1 = x^2 + (\dot{x} + ax + bx^2)^2$.
Return	
End	

Table 3.16. Numerical Results after Four Iterations for the First-Order Nonlinear System, Newton–Raphson Method

t	x	$\dot{x}$	y
0.0	1.000000	−1.461979	−0.3619785
0.1	0.8647847	−1.248209	−0.3086389
0.2	0.7492893	−1.066554	−0.2611212
0.3	0.6505893	−0.9114714	−0.2185554
0.4	0.5662597	−0.7784751	−0.1801505
0.5	0.4942811	−0.6638957	−0.1451833
0.6	0.4329682	−0.5647021	−0.1129877
0.7	0.3809126	−0.4783660	−0.08294396
0.8	0.3369378	−0.4027584	−0.05446797
0.9	0.3000639	−0.3360688	−0.02700109
1.0	0.2694788	−0.2767406	0.1764856 $E - 6$

main program. The Euler–Lagrange equation is

$$F_x - \frac{d}{dt} F_{\dot{x}} = 0, \tag{3.187}$$

which yields

$$2x + 2(\dot{x} + ax + bx^2)(a + 2bx) - 2(\ddot{x} + a\dot{x} + 2bx\dot{x}) = 0. \tag{3.188}$$

Solving for $\ddot{x}$, differentiating with respect to s, and expressing in terms of first-order differential equations as in equations (3.56)–(3.59) yields

$$\dot{x}_1 = x_2, \qquad x_1(0) = 1, \tag{3.189}$$

$$\dot{x}_2 = f_1, \qquad x_2(0) = -1, \tag{3.190}$$

$$\dot{x}_3 = x_4, \qquad x_3(0) = 0, \tag{3.191}$$

$$\dot{x}_4 = f_2, \qquad x_4(0) = 1, \tag{3.192}$$

where

$$f_1 = \ddot{x} = x - (a\dot{x} + 2bx\dot{x}) + (\dot{x} + ax + bx^2)(a + 2bx), \tag{3.193}$$

$$\begin{aligned} f_2 = \ddot{x}_s = x_s - (a\dot{x}_s + 2bx_s\dot{x} + 2bx\dot{x}_s) + (\dot{x} + ax + bx^2)(2bx_s) \\ + (\dot{x}_s + ax_s + 2bxx_s)(a + 2bx). \end{aligned} \tag{3.194}$$

The Newton–Raphson equation is obtained by expanding the boundary condition

$$\left.\frac{\partial F}{\partial \dot{x}}\right|_{t=T} = 0 \tag{3.195}$$

or

$$\dot{x}(T) + ax(T) + bx^2(T) = 0 \tag{3.196}$$

in a Taylor series and truncating after the first-order terms

$$\begin{aligned} &\dot{x}(T) + ax(T) + bx^2(T) \\ &\qquad + (s_{k+1} - s_k)[\dot{x}_s(T) + ax_s(T) + 2bx(T)x_s(T)] = 0 \end{aligned} \tag{3.197}$$

Solving for the s_{k+1} approximation yields the iteration equation

$$s_{k+1} = s_k - \frac{\dot{x}(T) + ax(T) + bx^2(T)}{\dot{x}_s(T) + ax_s(T) + 2bx(T)x_s(T)}. \tag{3.198}$$

Comparing the numerical results using this method with the numerical results using the automatic method, it can be shown that the results are in agreement to at least six digits.

Various other differential constraints have been tried using the automatic method, e.g., linear system $\dot{x} = -x + y$ and nonlinear system $\dot{x} = -x^2 + y$. The results were compared with the results in Reference 1. In all cases at least six- or seven-digit accuracies were obtained.

3.4.1.3. Second-Order Nonlinear Systems

The Newton–Raphson method with the Pontryagin maximum principle equations, as described in Section 3.2.1.2, is used in this section to obtain the numerical solution for a second-order system (Reference 15).

Consider minimizing the cost functional

$$J = \int_0^T (x_1^2 + y^2)\, dt \tag{3.199}$$

subject to the differential and initial constraints

$$\dot{x}_1 = x_2, \qquad x_1(0) = 1, \tag{3.200}$$

$$\dot{x}_2 = -b_1 x_1 - b_2 x_1^3 + y, \qquad x_2(0) = 0. \tag{3.201}$$

Equations (3.200) and (3.201) are the equations of a mass attached to a nonlinear spring. The Hamiltonian function is

$$H_1 = x_1^2 + y^2 + p_1 x_2 + p_2(-b_1 x_1 - b_2 x_1^3 + y). \tag{3.202}$$

The control that minimizes H_1 is

$$\frac{\partial H_1}{\partial y} = 2y + p_2 = 0, \tag{3.203}$$

$$y = -\tfrac{1}{2} p_2. \tag{3.204}$$

Substituting from equation (3.204) into equations (3.199)–(3.201) we obtain the integrand F and g_1 and g_2 as follows:

$$F = x_1^2 + \tfrac{1}{4} p_2^2, \tag{3.205}$$

$$g_1 = x_2, \qquad x_1(0) = 1, \tag{3.206}$$

$$g_2 = -b_1 x_1 - b_2 x_1^3 - \tfrac{1}{2} p_2, \qquad x_2(0) = 0. \tag{3.207}$$

Using the automatic method of solution, the only inputs required from the user for this example are the integrand, F, differential constraints, g_1 and g_2, the initial conditions, and the terminal time, T. The initial conditions and the terminal time are specified in the main program. The integrand, F, and the differential constraints, g_1 and g_2, are input in subroutine INPUT.

The automatic solution was obtained for the following conditions:

$$b_1 = 1, \qquad b_2 = 4, \qquad T = 1, \tag{3.208}$$

$$p_1(0) = 0, \qquad p_2(0) = 0. \tag{3.209}$$

Table 3.17 gives the FORTRAN listing of the INPUT subroutine. The integrand is represented by the vector **F**1 and the differential constraints by the vectors **G**1 and **G**2. The other variables and vectors are as defined in Table 3.6. The vector **D** is used to represent as needed the constant, 1/4 or 1/2, or the coefficient, b_1 or b_2. The vectors **E**1, **E**2, ..., **E**9 represent the intermediate variables. The vectors consist of 18 components since the last three components in Table 3.5 are not required. The complete program listing is given in Section 3.5.2.

A fourth-order Runge–Kutta integration method was used with grid intervals equal to 1/100. Table 3.18 shows the numerical results after the second iteration.

The numerical results were compared with the results obtained by solving for the Pontryagin and Newton–Raphson equations by hand, and substituting these equations in place of the INPUT subroutine. Application of Pontryagin's maximum principle to equations (3.199)–(3.201) yields

$$\dot{x}_1 = x_2, \qquad x_1(0) = 1, \tag{3.210}$$

$$\dot{x}_2 = -b_1 x_1 - b_2 x_1^3 - \tfrac{1}{2} p_2, \qquad x_2(0) = 0, \tag{3.211}$$

$$\dot{p}_1 = -2x_1 + b_1 p_2 + 3b_2 x_1^2 p_2, \qquad p_1(0) = c_1 \tag{3.212}$$

$$\dot{p}_2 = -p_1, \qquad p_2(0) = c_2 \tag{3.213}$$

Differentiating equations (3.210)–(3.213) with respect to c_1 and c_2 yields

$$\dot{x}_{1c_1} = x_{2c_1}, \qquad x_{1c_1}(0) = 0, \tag{3.214}$$

$$\dot{x}_{2c_1} = -b_1 x_{1c_1} - 3b_2 x_1^2 x_{1c_1} - \tfrac{1}{2} p_{2c_1}, \qquad x_{2c_1}(0) = 0, \tag{3.215}$$

$$\dot{p}_{1c_1} = -2x_{1c_1} + b_1 p_{2c_1} + 6b_2 x_1 x_{1c_1} p_2 + 3b_2 x_1^2 p_{2c_1}, \qquad p_{1c_1}(0) = 1, \tag{3.216}$$

$$\dot{p}_{2c_1} = -p_{1c_1}, \qquad p_{2c_1}(0) = 0, \tag{3.217}$$

and

$$\dot{x}_{1c_2} = x_{2c_2}, \qquad x_{1c_2}(0) = 0, \tag{3.218}$$

$$\dot{x}_{2c_2} = -b_1 x_{1c_2} - 3b_2 x_1^2 x_{1c_2} - \tfrac{1}{2} p_{2c_2}, \qquad x_{2c_2}(0) = 0, \tag{3.219}$$

$$\dot{p}_{1c_2} = -2x_{1c_2} + b_1 p_{2c_2} + 6b_2 x_1 x_{1c_2} p_2 + 3b_2 x_1^2 p_{2c_2}, \qquad p_{1c_2}(0) = 0, \tag{3.220}$$

$$\dot{p}_{2c_2} = -p_{1c_2}, \qquad p_{2c_2}(0) = 1. \tag{3.221}$$

Table 3.17. FORTRAN Listing of the INPUT Subroutine for the Second-Order Nonlinear System, Newton–Raphson Method

Listing	Purpose
Subroutine INPUT(X,CON,A1, A2,A3,A4,A5,F1,G1,G2)	
Dimension X(5),A1(18),A2(18), A3(18),A4(18),A5(18)	
Dimension D(18),F1(18), G1(18),G2(18)	
Dimension E1(18),E2(18), E3(18),E4(18),E5(18),E6(18)	
Dimension E7(18),E8(18), E9(18)	
Call LIN(X,A1,A2,A3,A4,A5)	Defines the vectors corresponding to t, x_1, x_2, p_1, and p_2.
Call MULT(A2,A2,E1)	Multiplies x_1 times x_1 to form $E1 = x_1^2$.
Call CONST(CON,D)	Defines the vector D corresponding to constant $= 1/4$.
Call MULT(A5,A5,E2)	Multiplies p_2 times p_2 to form $E2 = p_2^2$.
Call MULT(D,E2,E3)	Multiplies $1/4$ times p_2^2 to form $E3 = (1/4)p_2^2$.
Call ADD(E1,E3,F1)	Forms the sum $F1 = x_1^2 + \frac{1}{4}p_2^2$.
Call CONST(1.,D)	Defines the vector D corresponding to constant $= 1$.
Call MULT(D,A3,G1)	Multiplies 1 times x_2 to form $G1 = x_2$.
Call CONST(−.5,D)	Defines the vector D corresponding to constant $= -1/2$.
Call MULT(D,A5,E4)	Multiplies $-1/2$ times p_2 to form $E4 = -\frac{1}{2}p_2$.
B1 = −1.0	Defines the coefficient $b_1 = 1$.
B2 = −4.0	Defines the coefficient $b_2 = 4$.
Call CONST(B1,D)	Defines the vector D corresponding to coefficient $b_1 = 1$.
Call MULT(D,A2,E5)	Multiplies $-b_1$ times x_1 to form $E5 = -b_1x_1$.
Call ADD(E4,E5,E6)	Forms the sum $E6 = -b_1x_1 - \frac{1}{2}p_2$.
Call CONST(B2,D)	Defines the vector D corresponding to coefficient $b_2 = 4$.
Call MULT(A2,A2,E7)	Multiplies x_1 times x_1 to form $E7 = x_1^2$.
Call MULT(E7,A2,E8)	Multiplies x_1^2 times x_1 to form $E8 = x_1^3$.
Call MULT(D,E8,E9)	Multiplies $-b_2$ times x_1^3 to form $E9 = -b_2x_1^3$.
Call ADD(E6,E9,G2)	Forms the sum $G2 = -b_1x_1 - b_2x_1^3 - \frac{1}{2}p_2$.
Return	
End	

Table 3.18. Numerical Results after the Second Iteration for the Second-Order Nonlinear System, Newton–Raphson Method

t	x_1	x_2
0.0	1.0	0.0
0.1	0.9748863	−0.4964506
0.2	0.9028140	−0.9305701
0.3	0.7921403	−1.264718
0.4	0.6534822	−1.491325
0.5	0.4969368	−1.626423
0.6	0.3303637	−1.696475
0.7	0.1589621	−1.726623
0.8	−0.0142007	−1.733571
0.9	−0.1871644	−1.722473
1.0	−0.3578697	−1.686222

Using equations (3.210)–(3.221) instead of the table method yields results that agree with the results in Table 3.18 to at least six or seven digits. The advantage of the table method, of course, is that equations (3.210)–(3.221) do not have to be derived and that the same program can be used for different cost functionals and constraints. Both programs can be used, however, for different values of b_1 and b_2.

Additional numerical results were obtained for $b_1 = 1$, $b_2 = 0$ and $b_1 = 0$, $b_2 = 0$. The two-point boundary value problems are linear for these two cases, so that the solutions are obtained in one iteration. Again using the automatic method and the method using equations (3.210)–(3.221) the results agree to at least six or seven digits. The results for $b_1 = 0$, $b_2 = 0$ also agree with the results given in Reference 1.

3.4.1.4. *First-Order Systems with Integral Constraints*

Consider the isoperimetric problem of finding the control that minimizes the cost functional (Reference 15)

$$J_1 = \int_0^T x^2 \, dt \tag{3.222}$$

subject to the integral constraint

$$E = \int_0^T y^2 \, dt \tag{3.223}$$

and differential constraint

$$\dot{x} = -ax + y, \qquad x(0) = x_0. \tag{3.224}$$

The problem can be solved by minimizing the integral

$$J = \int_0^T [x^2 + q(y^2 - E/T)]\, dt, \tag{3.225}$$

where q is the Lagrange multiplier and is equal to a constant.

It was shown in Reference 1 that this problem can be solved by adjoining the differential constraint

$$\dot{q} = 0 \tag{3.226}$$

with the unknown initial condition, $q(0)$. The Hamiltonian function is then

$$H = x^2 + qy^2 + p_1(-ax + y) + p_2 \times 0, \tag{3.227}$$

where $p_2(t)$ is the augmented costate variable due to differential constraint (3.226). The two-point boundary value equations and associated boundary conditions are

$$\dot{x} = -ax + \frac{1}{2q} p_1, \qquad x(0) = x_0, \tag{3.228}$$

$$\dot{q} = 0 \qquad p_2(0) = 0, \tag{3.229}$$

$$\dot{p}_1 = -2x + ap_1, \qquad p_1(T) = 0, \tag{3.230}$$

$$\dot{p}_2 = -\frac{1}{4q^2} p_1^2, \qquad p_2(T) = -E, \tag{3.231}$$

where the optimal control is

$$y = -\frac{1}{2q} p_1. \tag{3.232}$$

The unknown initial conditions are

$$p_1(0) = c_1, \qquad q(0) = c_2. \tag{3.233}$$

The Newton–Raphson iteration equations (3.81) and (3.82) still apply provided that $p_2(T, c_1, c_2)$ is replaced by

$$p_2(T, c_1, c_2) = p_2(T, c_1, c_2) + E \tag{3.234}$$

and that

$$p_{2c_2}(0) = 0, \qquad q_{c_2}(0) = 1. \tag{3.235}$$

The integrand F and differential constraints g_1 and g_2 for the isoperimetric problem given by equations (3.222)–(3.224) are then

$$F = x_2 + \frac{1}{4q} p_1^2, \tag{3.236}$$

$$g_1 = -ax - \frac{1}{2q} p_1, \qquad x_1(0) = x_0, \tag{3.237}$$

$$g_2 = 0, \qquad q(0) = c_2. \tag{3.238}$$

The program for the automatic solution was utilized for the following conditions:

$$a = 1, \qquad T = 1, \qquad E = 0.0393639, \tag{3.239}$$

$$q(0) = 2, \qquad p_1(0) = 1.4. \tag{3.240}$$

It is known that $q(0)$ should be equal to one for the above conditions.

The FORTRAN listing of the INPUT subroutine is given in Table 3.19. Table 3.20 shows the convergence of q. The value of q reaches its steady-state value in five iterations. The state x as a function of time after five iterations is given in Table 3.21. The numerical results agree with the analytical solution given in Reference 1 to at least six or seven digits.

3.4.2. Gradient Method

Examples of optimal control problems using the gradient method of solution are given in this section. The gradient method, as described in Section 3.2.2, has the advantage that the control function does not have to be solved explicitly as a function of the state and costate variables. If the optimal control is obtained with the evaluation of the minimum cost functional, then both the Hamiltonian and the integrand of the cost functional must be input in the INPUT subroutine. If the optimal control is obtained using the Newton–Raphson recurrence relation then only the Hamiltonian function is required. Another advantage of using the gradient method is that only the first derivatives are required, and if the Newton–Raphson recurrence relation is used, only an additional single second-order derivative, H_{yy}, is required. Thus the dimensions of the vectors used in the automatic solution method are reduced considerably.

Table 3.19. FORTRAN Listing of the INPUT Subroutine for the First-Order System with the Integral Constraint, Newton–Raphson Method

Listing	Purpose
Subroutine INPUT(X,CON,A1, A2,A3,A4,A5,F1,G1,G2)	
Dimension X(5),A1(18),A2(18), A3(18),A4(18),A5(18)	
Dimension D(18),F1(18), G1(18),G2(18)	
Dimension E1(18),E2(18), E3(18),E4(18)	
Dimension E5(18),E6(18), E7(18)	
Call LIN(X,A1,A2,A3,A4,A5)	Defines the vectors corresponding to t, x, q, p_1, and p_2.
Call MULT(A2,A2,E1)	Multiplies x times x to form $E1 = x^2$.
Call CONST(CON,D)	Defines the vector D corresponding to constant $= 1/4$.
Call MULT(A4,A4,E2)	Multiplies p_1 times p_1 to form $E2 = p_1^2$.
Call MULT(D,E2,E3)	Multiplies $1/4$ times p_1^2 to form $E3 = (1/4)p_1^2$.
Call DIV(A3,E3,E4)	Divides E_3 by q to form $E4 = (1/4q)p_1^2$.
Call ADD(E1,E4,F1)	Forms the sum $F1 = x^2 + (1/4q)p_1^2$.
Call CONST(−1.0,D)	Defines the vector D corresponding to coefficient $a = 1$.
Call MULT(D,A2,E5)	Multiplies $-a$ times x to form $E5 = -ax$.
Call CONST(−.5,D)	Defines the vector D corresponding to constant $= -1/2$.
Call MULT(D,A4,E6)	Multiplies $-1/2$ times p_1 to form $E6 = -\frac{1}{2}p_1$.
Call DIV(A3,E6,E7)	Divides $E6$ by q to form $E7 = -(1/2q)p_1$.
Call ADD(E5,E7,G1)	Forms the sum $G1 = -ax - (1/2q)p_1$.
Call CONST(0.0,D)	Defines the vector D corresponding to constant $= 0$.
Call MULT(D,A5,G2)	Multiplies 0 times p_2 to form $G2 = 0$.
Return	
End	

Table 3.20. Convergence of q for the First-Order System with the Integral Constraint, Newton–Raphson Method

Iteration number	q
0	2.0
1	1.110578
2	0.9600788
3	0.9974809
4	0.9999906
5	1.000001

3.4.2.1. First-Order Nonlinear Systems

Three examples of using the gradient method with the evaluation of the minimum cost functional, as described in Section 3.2.2.1, are given (Reference 16).

Example 3.1. Consider minimizing the cost functional given in Section 3.4.1.2

$$J = \int_0^T (x^2 + y^2)\, dt \tag{3.241}$$

Table 3.21. Numerical Results after the Fifth Iteration for the First-Order System with the Integral Constraint, Newton–Raphson Method

t	x	q
0.0	1.0	1.000001
0.1	0.8709724	1.000001
0.2	0.7593934	1.000001
0.3	0.6630275	1.000001
0.4	0.5799443	1.000001
0.5	0.5084793	1.000001
0.6	0.4472008	1.000001
0.7	0.3948813	1.000001
0.8	0.3504726	1.000001
0.9	0.3130850	1.000001
1.0	0.2819696	1.000001

subject to the differential constraint

$$\dot{x} = -ax - bx^2 + y, \qquad x(0) = x_0. \tag{3.242}$$

The integrand and the Hamiltonian function are

$$F = x^2 + y^2, \tag{3.243}$$

$$H = x^2 + y^2 + p(-ax - bx^2 + y). \tag{3.244}$$

The FORTRAN program for the automatic solution was utilized for the following conditions:

$$a = 1, \qquad b = 0.1, \qquad T = 1. \tag{3.245}$$

The constraint equation and initial condition are

$$\dot{x} = -x - 0.1x^2 + y, \qquad x(0) = 1. \tag{3.246}$$

The initial guess of the solution for the optimal control and the initial value of K for all the examples were

$$y(t) = 0, \qquad 0 \leq t \leq T, \qquad K_{\text{initial}} = 1.0. \tag{3.247}$$

The initial value of K was set equal to 1 such that the four values of K_i for the first iteration were 1/2, 1, 2, and 4.

The only inputs required from the user are the integrand, F, the Hamiltonian, H, the initial conditions, and the terminal time, T. The initial conditions and the terminal time are specified in the main program. The integrand, F, and the Hamiltonian, H, are input in Subroutine INPUT. Table 3.22 gives the FORTRAN listing of the INPUT subroutine. The Hamiltonian is represented by the vector **HA** and the integrand by the vector **F**. The other variables and vectors are as defined in Table 3.9. The vector **D** is used to represent as needed the coefficient a or b. The vectors **E**1, . . . , **E**5 represent the intermediate variables. The vectors consist of five components as given in Table 3.8.

Table 3.23 shows the convergence of the terminal value of the state, $x(T)$, and the cost as a function of the iteration number. Table 3.24 shows the numerical solutions, $x(t)$ and $y(t)$, as a function of time after the fifth iteration. The results agree with the Newton–Raphson solution given in Section 3.4.1.2 to approximately four digits.

The numerical results were also compared with the results obtained by solving for the partial derivatives by hand and substituting these equa-

Table 3.22. FORTRAN Listing of the INPUT Subroutine for the First-Order Nonlinear System, Gradient Method

Listing	Purpose
Subroutine INPUT(X,HA,F)	
Dimension A1(5),A2(5),A3(5), A4(5)	
Dimension X(4),HA(5),D(5)	
Dimension E1(5),E2(5),E3(5), E4(5),E5(5),F(5)	
Call LIN(X,A1,A2,A3,A4)	Defines the vectors corresponding to t, x, p, and y.
B1 = −1	Defines the coefficient $a = 1$.
B2 = −0.1	Defines the coefficient $b = 0.1$.
Call CONST(B1,D)	Defines the vector D corresponding to coefficient a.
Call MULT(D,A2,E1)	Multiplies $-a$ times x to form $E1 = -ax$.
Call MULT(A2,A2,E2)	Multiplies x times x to form $E2 = x^2$.
Call CONST(B2,D)	Defines the vector D corresponding to coefficient b.
Call MULT(D,E2,E3)	Multiplies $-b$ times x^2 to form $E3 = -bx^2$.
Call ADD(E1,E3,E4)	Forms the sum $E4 = -ax - bx^2$.
Call ADD(A4,E4,E5)	Forms the sum $E5 = \dot{x} = -ax - bx^2 + y$.
Call MULT(A3,E5,E4)	Multiplies p times $\dot{x}$ to form $E4 = p(-ax - bx^2 + y)$.
Call MULT(A4,A4,E5)	Multiplies y times y to form $E5 = y^2$.
Call ADD(E2,E5,F)	Forms the sum $F = x^2 + y^2$.
Call ADD(F,E4,HA)	Forms the sum $\text{HA} = x^2 + y^2 + p(-ax - bx^2 + y)$.
Return	
End	

tions in place of the INPUT subroutine. For this particular example, Pontryagin's maximum principle yields

$$\dot{x} = H_p = f(t, x, y) = -ax - bx^2 + y, \qquad x(0) = 1, \tag{3.248}$$

$$\dot{p} = -H_x = -2x + ap + 2bxp, \qquad p(1) = 0, \tag{3.249}$$

$$H_y = 2y + p. \tag{3.250}$$

The numerical results were the same as for the automatic method of solution.

Table 3.23. Convergence of the Optimal Control Problem Examples, First-Order Nonlinear System, Gradient Method

Iteration number	Example 1 $\dot{x} = -x - 0.1x^2 + y$ $x(0) = 1, \quad T = 1$		Example 2 $\dot{x} = -x^2 + y$ $x(0) = 10, \quad T = 1$		Example 3 $\dot{x} = -y$ $x(0) = 10, \quad T = 0.1$	
	$x(T)$	Cost	$x(T)$	Cost	$x(T)$	Cost
0	0.3460076	—	0.9090910	—	10.00000	—
1	0.2542664	0.3708872	0.8332673	8.971876	9.950000	9.966800
2	0.2722690	0.3697418	0.8370252	8.971778	9.950207	9.966799
3	0.2689736	0.3697052	0.8368574	8.971778		
4	0.2695669	0.3697040	0.8368866	8.971777		
5	0.2694605	0.3697040	0.8368404	8.971778		

Table 3.24. Numerical Solutions, First-Order Nonlinear System, Gradient Method

t	Example 1 Fifth iteration		Example 2 Fifth iteration		Example 3[a] Second iteration	
	$x(t)$	$y(t)$	$x(t)$	$y(t)$	$x(t)$	$y(t)$
0.0	1.000000	−0.3620199	10.00000	−0.4952823	10.00000	−0.9966750
0.1	0.8647802	−0.3086828	4.971543	−0.4811673	9.990533	−0.8967225
0.2	0.7492806	−0.2611662	3.287593	−0.4579327	9.982065	−0.7968600
0.3	0.6505770	−0.2185998	2.440042	−0.4259962	9.974595	−0.6970775
0.4	0.5662443	−0.1801926	1.928318	−0.3858791	9.968123	−0.5973650
0.5	0.4942632	−0.1452215	1.585959	−0.3381490	9.962648	−0.4977125
0.6	0.4329486	−0.1130205	1.341890	−0.2833550	9.958169	−0.3981100
0.7	0.3808920	−0.08296997	1.160782	−0.2219592	9.954685	−0.2985475
0.8	0.3369171	−0.05448605	1.023187	−0.1542669	9.952198	−0.1990150
0.9	0.3000440	−0.02701043	0.9175947	−0.08035531	9.950705	−0.09950250
1.0	0.2694605	0.0000000	0.8368404	0.0000000	9.950207	0.0000000

[a] Multiply time, t, in the first column by 0.1 since the terminal time $T = 0.1$ for example 3.

Linear interpolation was used between the grid intervals for both the state, $x(t)$, and the control, $y(t)$. Without linear interpolation the results agree with the Newton–Raphson method of solution to only approximately three digits instead of four.

Example 3.2. Example 2 is the same as 1 except that

$$a = 0, \qquad b = 1, \qquad T = 1. \tag{3.251}$$

The constraint equation and initial condition are

$$\dot{x} = -x^2 + y, \qquad x(0) = 10. \tag{3.252}$$

This problem is considered by Sage on page 313 in Reference 4. As pointed out in that reference, for $K_i = 0.5$, the trajectories have almost converged to their correct values after just one iteration. Note, however, that the convergence is slow after the first iteration. As shown in Table 3.23, the last three digits of the terminal value of the state, $x(T)$, have not reached steady state after five iterations even though the cost is not changing (except for the last digit). Table 3.24 shows the numerical solutions, $x(t)$ and $y(t)$, as a function of time after the fifth iteration. The results agree with the Newton–Raphson method of solution, obtained by using the program described in Section 3.4.1.2, to approximately four digits.

Example 3.3. Example 3 is the same as 1 except that

$$a = 0, \qquad b = 0, \qquad T = 0.1. \tag{3.253}$$

The constraint equation and initial condition are

$$\dot{x} = y, \qquad x(0) = 10. \tag{3.254}$$

Table 3.23 shows the convergence of the terminal value of the state, $x(T)$, and the cost as a function of the iteration number. Table 3.24 shows the numerical solutions, $x(t)$ and $y(t)$, as a function of time after the second iteration. The analytical solution for this linear problem is

$$x(t) = 10(\cosh t - \tanh T \sinh t). \tag{3.255}$$

The numerical solution for this example agrees with the analytical solution exactly.

3.4.2.2. Second-Order Nonlinear Systems

The solution of the second-order system example given in Section 4.4.1.3 is obtained in this section using the gradient method with the Newton–Raphson recurrence relation described in Section 3.2.2.2. Some results for first-order systems are also given.

Consider minimizing the cost functional (References 17, 18)

$$J = \int_0^T (x_1^2 + y^2)\, dt \tag{3.256}$$

subject to the second-order nonlinear differential constraint

$$\dot{x}_1 = x_2, \qquad x_1(0) = 1, \tag{3.257}$$

$$\dot{x}_2 = -b_1 x_1 - b_2 x_1^3 + y, \qquad x_2(0) = 0. \tag{3.258}$$

The Hamiltonian function is

$$H = x_1^2 + y^2 + p_1 x_2 + p_2(-b_1 x_1 - b_2 x_1^3 + y). \tag{3.259}$$

The program for the automatic solution was utilized for the following conditions:

$$b_1 = 1, \qquad b_2 = 4, \qquad T = 1. \tag{3.260}$$

The initial guess of the solution for the optimal control was

$$y(t) = 0, \qquad 0 \leq t \leq 1. \tag{3.261}$$

The only inputs required from the user are the Hamiltonian function, H, the initial conditions, and the terminal time T. The Hamiltonian, H, is input in subroutine INPUT. Table 3.25 gives the FORTRAN listing of the INPUT subroutine.

The Hamiltonian is represented by the vector **HA**. The other variables and vectors are as defined in Table 3.12 for an nth-order system where $n = 2$. The subroutine is dimensioned to handle any order system from 1 to 3. The example in this case is for a second-order system. The vector **D** is used to represent as needed the coefficient b_1 or b_2. The vectors **E**1, ..., **E**5 represent the intermediate variables. The vector consist of $2n + 4 = 8$ components as shown in Table 3.11. The subroutine in Table 3.25 is dimensioned for 10 components; however, only eight are used.

Table 3.26 shows the convergence of the terminal values of the states, $x_1(T)$ and $x_2(T)$, as a function of the iteration number. The Newton–Raph-

Table 3.25. FORTRAN Listing of the INPUT Subroutine for the Second-Order Nonlinear System, Gradient Method

Listing	Purpose
Subroutine INPUT(X,HA)	
Common L,X1D,ID,L1,LV,IV,A	
Dimension A(8,10),T(10),Y(10)	
Dimension X1(10),X2(10),X3(10)	
Dimension P1(10),P2(10),P3(10)	
Dimension X(IV),HA(LV),F(10), D(10)	
Dimension E1(10),E2(10),E3(10), E4(10),E5(10)	
Call LIN(X)	Defines the $(2n + 2) \times (2n + 4)$ matrix A corresponding to t, x_1, x_2, p_1, p_2, and y.
Do 325 I = 1,LV T(I) = A(1,I) Go to(1,2,3)ID	Do-loop forms the vectors from the rows of the matrix A for LV $= 2n + 4 = 8$ and ID $= n = 2$.
3 X3(I) = A(4,I) P3(I) = A(4+ID,I)	
2 X2(I) = A(3,I) P2(I) = A(3+ID,I)	
1 X1(I) = A(2,I) P1(I) = A(2+ID,I)	
325 Y(I) = A(IV,I)	
B1 = −1.	Defines the coefficient $b_1 = 1$.
B2 = −4.	Defines the coefficient $b_2 = 4$.
Call CONST(B1,D)	Defines the vector D corresponding to the coefficient b_1.
Call MULT(D,X1,E1)	Multiplies $-b_1$ times x_1 to form $E1 = -b_1x_1$.
Call MULT(X1,X1,E2)	Multiplies x_1 times x_1 to form $E2 = x_1^2$.
Call MULT(X1,E2,E3)	Multiplies x_1 times x_1^2 to form $E3 = x_1^3$.
Call CONST(B2,D)	Defines the vector D corresponding to the coefficient b_2.
Call MULT(D,E3,E4)	Multiplies $-b_2$ times x_1^3 to form $E4 = -b_2x_1^3$.
Call ADD(E1,E4,E5)	Forms the sum $E5 = -b_1x_1 - b_2x_1^3$.
Call ADD(E5,Y,E1)	Forms the sum $E1 = \dot{x}_2 = -b_1x_1 - b_2x_1^3 + y$.
Call MULT(E1,P2,E5)	Multiplies p_2 times $\dot{x}_2$ to form $E5 = p_2(-b_1x_1 - b_2x_1^3 + y)$.

Table 3.25. (*Continued*)

Listing	Purpose
Call MULT(P1,X2,E1)	Multiplies p_1 times x_2 to form $E1 = p_1x_2$.
Call ADD(E5,E1,E4)	Forms the sum $E4 = p_1x_2 + p_2(-b_1x_1 - b_2x_1^3 + y)$.
Call MULT(Y,Y,E5)	Multiplies y times y to form $E5 = y^2$.
Call ADD(E2,E5,F)	Forms the sum $F = x_1^2 + y^2$.
Call ADD(F,E4,HA)	Forms the sum $\text{HA} = x_1^2 + y^2 + p_1x_2 + p_2(-b_1x_1 - b_2x_1^3 + y)$.
Return	
End	

son recurrence relation, equation (3.134), was used. Table 3.27 shows the numerical solutions, $x_1(t)$, $x_2(t)$, and $y(t)$, as a function of time after the fifth iteration. The results agree with the Newton–Raphson method of solution in Section 3.4.1.3 to approximately five digits.

To further demonstrate the advantages of using the Newton–Raphson recurrence relation with the automatic calculation of the derivatives, the recurrence relation was applied to the first-order system example given in Section 3.4.2.1:

$$J = \int_0^T (x^2 + y^2)\, dt, \tag{3.262}$$

$$\dot{x} = -x - 0.1x^2 + y, \qquad x(0) = 1. \tag{3.263}$$

Table 3.26. Convergence of the Second-Order Nonlinear Optimal Control Problem, Gradient Method

Iteration number	$x_1(T)$	$x_2(T)$
0	−0.3523335	−1.686725
1	−0.3581911	−1.686541
2	−0.3578567	−1.686208
3	−0.3578714	−1.686224
4	−0.3578708	−1.686223
5	−0.3578708	−1.686223

Table 3.27. Numerical Solution after the Fifth Iteration for the Second-Order Nonlinear System, Gradient Method

t	$x_1(t)$	$x_2(t)$	$y(t)$
0.0	1.0	0.0	$-0.8644429\,E-1$
0.1	0.9748863	-0.4964505	$-0.5833226\,E-1$
0.2	0.9028140	-0.9305702	$-0.3266744\,E-1$
0.3	0.7921402	-1.264719	$-0.1236860\,E-1$
0.4	0.6534821	-1.491326	$0.1201212\,E-2$
0.5	0.4969366	-1.626424	$0.8250635\,E-2$
0.6	0.3303633	-1.696476	$0.1004007\,E-1$
0.7	0.1589615	-1.726625	$0.8299332\,E-2$
0.8	$-0.1420149\,E-1$	-1.733573	$0.4856373\,E-2$
0.9	-0.1871654	-1.722475	$0.1503933\,E-2$
1.0	-0.3578708	-1.686223	0.0

The gradient method was used in that section with the value of K selected such that the smallest J_i^{N+1} is obtained for the four values of K_i chosen, as described in Section 3.2.2.1. Numerical results identical to those in Section 3.4.2.1 were obtained. The CPU (central processing unit) time was reduced from approximately 15 to 5 sec. The second derivative of the Hamiltonian, H_{yy}, is equal to the constant, 2, in this example.

Replacing system equation (3.263) by

$$\dot{x} = -x - 0.1x^2 + e^y \tag{3.264}$$

the second derivative of the Hamiltonian, H_{yy}, is no longer a constant, but varies from approximately 3 to 2 during the course of the trajectory for each iteration. The final value of $x(T)$ after five iterations is 0.8402830. On the other hand, using the constant K with $K_{\text{initial}} = 0.5$, the K_i selected for minimum J_i^{N+1} is 0.25 for the first four iterations and 0.5 for the fifth iteration. The final value of $x(T)$ after five iterations is 0.8400046. It is assumed that the former value of $x(T)$ is more nearly equal to the true value.

3.4.2.3. Third-Order Nonlinear Systems

An example of a third-order system (Reference 18) is given in this section using the gradient method with the Newton–Raphson recurrence relation described in Section 3.2.2.2.

Consider the unconstrained brachistochrone problem given on page 323 in Reference 4. The problem is to find the path of a particle falling in a constant gravitational field, g, such that the final value of the horizontal coordinate, $x_1(T)$, is maximized. The vertical coordinate is x_2. The terminal time, T, and the initial speed, $x_3(0)$ are given.

The system equations are

$$\dot{x}_1 = x_3 \cos y, \qquad x_1(0) = 0, \tag{3.265}$$

$$\dot{x}_2 = x_3 \sin y, \qquad x_2(0) = 0, \tag{3.266}$$

$$\dot{x}_3 = g \sin y, \qquad x_3(0) = 0.07195, \tag{3.267}$$

$$g = 1, \qquad T = 1.7, \tag{3.268}$$

with cost function

$$J = -x_1(T). \tag{3.269}$$

The Hamiltonian function is given by

$$H = p_1 x_3 \cos y + p_2 x_3 \sin y + p_3 \sin y, \tag{3.270}$$

where p_1, p_2, and p_3 are the costate variables. The terminal conditions for the costate variables are

$$p_1(T) = \quad 1, \qquad p_2(T) = 0, \qquad p_3(T) = 0. \tag{3.271}$$

The program for the automatic solution was utilized with the initial guess of the solution for the optimal control given by

$$y(t) = \pi/6, \qquad 0 \leq t \leq T. \tag{3.272}$$

The only inputs required from the user are the Hamiltonian function, H, the initial conditions and the terminal time T. The Hamiltonian is input in subroutine INPUT. Table 3.28 gives the FORTRAN listing of the INPUT subroutine. The complete program listing is given in Section 3.5.3.

The Hamiltonian is represented by the vector **HA**. The other variables and vectors are as defined in Table 3.12 for an nth-order system, where for the third-order system used as an example here, $n = 3$. The subroutine is dimensioned to handle any order system from 1 to 4. The vectors **E**1, . . . , **E**8, **D**, **D**1, **D**2, and **F** represent the intermediate variables; however, not all of them are required for this example. The number of vectors corresponding to the variables is equal to $2n + 2 = 8$. The number of com-

Table 3.28. FORTRAN Listing of the INPUT Subroutine for the Third-Order Nonlinear System, Gradient Method

Listing	Purpose
Subroutine INPUT(X,HA)	
Common L,X1D,ID,L1,LV,IV,A	
Dimension A(10,12),T(12),Y(12)	
Dimension X1(12),X2(12), X3(12),X4(12)	
Dimension P1(12),P2(12), P3(12),P4(12)	
Dimension X(IV),HA(LV), F(12),D(12)	
Dimension E1(12),E2(12),E3(12), E4(12),E5(12),E6(12)	
Dimension E7(12),E8(12), D1(12),D2(12)	
Call LIN(X)	Defines the $(2n+2) \times (2n+4)$ matrix A corresponding to t, x_1, x_2, x_3, p_1, p_2, p_3, and y.
Do 325 I = 1,LV	Do-loop forms the vectors from the rows of the matrix A for LV $= 2n + 4 = 10$ and ID $= n = 3$
T(I) = A(1,I)	
Go to(1,2,3,4)ID	
4 X4(I) = A(5,I)	
P4(I) = A(5+ID,I)	
3 X3(I) = A(4,I)	
P3(I) = A(4+ID,I)	
2 X2(I) = A(3,I)	
P2(I) = A(3+ID,I)	
1 X1(I) = A(2,I)	
P1(I) = A(2+ID,I)	
325 Y(I) = A(IV,I)	
Call SCOS(Y,E1)	Defines the function $E1 = \cos y$.
Call MULT(X3,E1,E2)	Multiplies x_3 times $\cos y$ to form $E2 = x_3 \cos y$.
Call MULT(P1,E2,E3)	Multiplies p_1 times $x_3 \cos y$ to form $E3 = p_1 x_3 \cos y$.
Call SSIN(Y,E1)	Defines the function $E1 = \sin y$.
Call MULT(X3,E1,E2)	Multiplies x_3 times $\sin y$ to form $E2 = x_3 \sin y$.
Call MULT(P2,E2,E4)	Multiplies p_2 times $x_3 \sin y$ to form $E4 = p_2 x_3 \sin y$.
Call ADD(E3,E4,E5)	Forms the sum $E5 = p_1 x_3 \cos y + p_2 x_3 \sin y$.
Call MULT(P3,E1,E2)	Multiplies p_3 times $\sin y$ to form $E2 = p_3 \sin y$.
Call ADD(E5,E2,HA)	Forms the sum HA $= p_1 x_3 \cos y + p_2 x_3 \sin y + p_3 \sin y$.
Return	
End	

Table 3.29. Numerical Solution after the Eighth Iteration for the Third-Order Nonlinear System, Gradient Method

t	$x_1(t)$	$x_2(t)$	$y(t)$
0.00	0.0	0.0	1.507033
0.34	0.0199929	0.0785826	1.205895
0.68	0.1147656	0.2407435	0.9046448
1.02	0.318535	0.4293704	0.6032325
1.36	0.6271264	0.5779684	0.3016638
1.70	0.9993763	0.6341072	0.0

ponents of the vectors is equal to $2n + 4 = 10$ as shown in Table 3.11. The subroutine in Table 3.28 is dimensioned for 12 components; however, only 10 are used for $n = 3$. Statement numbers 1–3 in Table 3.28 convert the rows of matrix A into the eight vectors **T**, **X1**, **X2**, **X3**, **P1**, **P2**, **P3**, and **Y**. The program can be dimensioned to handle an nth-order system, where $n \geq 1$, as described in Section 3.5.3.

Table 3.29 shows the numerical solutions for $x_1(t)$, $x_2(t)$, and $y(t)$ as a function of time after the eighth iteration. The results are accurate to approximately three digits. The analytical solution for the optimum path can be shown to be a cycloid given by the equations

$$x_1(t) = x_0 + a(\theta - \sin\theta), \tag{3.273}$$

$$x_2(t) = -\frac{x_3^2(0)}{2g} + a(1 - \cos\theta), \tag{3.274}$$

$$\theta(t) = \left(\frac{g}{a}\right)^{1/2} t + \theta_0, \tag{3.275}$$

where for the given example

$$a = \frac{1 - x_0}{\pi}, \tag{3.276}$$

$$x_0 = -0.00011, \tag{3.277}$$

$$\theta_0 = 0.1276. \tag{3.278}$$

At the terminal time, $t = T$

$$x_1(T) = 1, \qquad x_2(T) = 0.634, \qquad \theta(T) = \pi. \tag{3.279}$$

3.5. Program Listings

The FORTRAN program listings are given in this section for selected optimal control problem examples described in Section 3.4. Section 3.5.1 gives the program listing for the simplest problem in the calculus of variations using the Newton–Raphson method with the Euler–Lagrange equations (Reference 13). The program listing for second-order system optimal control problems and for first-order systems with integral constraints (Reference 15) is given in Section 3.5.2. The program uses the Newton–Raphson method with Pontryagin's maximum principle. The program listing for nth-order systems (Reference 18) using a third-order system example is given in Section 3.5.3. This program uses the gradient method with the Newton–Raphson recurrence relation. A brief description of the program is given with the listing in each section. The differential equations are integrated using a fourth-order Runge–Kutta integration method with grid intervals $\Delta t = 1/100$.

3.5.1. Simplest Problem in the Calculus of Variations

Method of solution: Newton–Raphson method using the Euler–Lagrange equations, Section 3.2.1.1.

Table 3.30. Definitions of the FORTRAN Variables and Vectors for the Given Example of the Simplest Problem in the Calculus of Variations

	Variable	FORTRAN variable or vector	Number of components	Program line number
Initial conditions	$\mathbf{x}(0)$	CI	4	130–160
Terminal condition	d	DO	1	170
Integration limits	a	X1I	1	180
	b	X1L	1	190
Integrand	$F = \dfrac{(1+\dot{x})^{1/2}}{x}$	F1	20	920
Integration grid interval size	Δt	H	1	3215
State variables[a]	$\mathbf{x}$	Z	4	3480
Derivatives to be integrated each iteration	$\dot{\mathbf{x}}$	DZ	4	3780–3825

[a] Where $\mathbf{x} = (x, \dot{x}, x_s, \dot{x}_s)^T$. Additional definitions of the variables and vectors are given in Tables 3.2 and 3.3.

Description of example and numerical results: Section 3.4.1.1.

Definitions of the FORTRAN variables and vectors for the given example: Table 3.30.

FORTRAN program: Listing 1.

FORTRAN Program Listing 1
Simplest Problem in the Calculus of Variations

```
100          DIMENSION A(20),B(20),C(20),F1(20)
110          DIMENSION CI(4),Z(4),DZ(4)
130          CI(1)=1.
140          CI(2)=.1
150          CI(3)=0.
160          CI(4)=1.0
170          DO=2.
180          X1I=1.
190          X1L=2.
210          X1D=X1L-X1I
220          CON=1.0
300          DO 150 N1=1,6
310          CALL MMULT(CI,1.,Z)
320          X1=X1I
330          CALL INTEG(X1,X2,X3,CON,X1D,A,B,C,F1,Z,DZ)
340    150   CI(2)=CI(2)-(Z(1)-DO)/Z(3)
350          END
850          SUNROUTINE INPUT(X1,X2,X3,CON,A,B,C,F1)
860          DIMENSION A(20),B(20),C(20),D(20),E(20)
865          DIMENSION G(20),S(20),F1(20)
870          CALL LIN(X1,X2,X3,A,B,C)
880          CALL CONST(CON,D)
890          CALL MULT(C,C,E)
900          CALL ADD(D,E,G)
910          CALL SR(G,S)
920          CALL DIV(B,S,F1)
930          RETURN
940          END
950          SUBROUTINE LIN(X1,X2,X3,A,B,C)
960          DIMENSION A(20),B(20),C(20)
970          DO 10 I=1,20
980          A(I)=0.0
990          B(I)=0.0
1000   10     C(I)=0.0
1010          A(1)=X1
1020          A(2)=1.0
1030          B(1)=X2
1040          B(3)=1.0
1050          C(1)=X3
1060          C(4)=1.0
1070          RETURN
1080          END
```

```
1090          SUBROUTINE CONST(CON,D)
1100          DIMENSION D(20)
1110          DO 15 I=1,20
1120    15    D(I)=0.0
1130          D(1)=CON
1140          RETURN
1150          END
1160          SUBROUTINE MULT(A,B,E)
1170          DIMENSION A(20),B(20),E(20)
1180          E(1)=A(1)*B(1)
1190          DO 15 I=1,3
2000    15    E(I+1)=A(I+1)*B(1)+A(1)*B(I+1)
2010          L=4
2020          DO 16 I=1,3
2030          DO 17 J=I,3
2040          L=L+1
2050    17    E(L)=A(L)*B(1)+A(I+1)*B(J+1)+A(J+1)*B(I+1)+A(1)*B(L)
2070    16    CONTINUE
2080          L=10
2090          IJ=4
2100          DO 18 I=1,3
2110          DO 19 J=I,3
2120          IJ=IJ+1
2130          DO 20 K=J,3
2140          IK=4
2150          DO 25 I1=1,3
2160          DO 26 K1=I1,3
2170          IK=IK+1
2180          IF(I1.EQ.I.AND.K1.EQ.K)GO TO 30
2190    26    CONTINUE
2200    25    CONTINUE
2210    30    JK=4
2220          DO 35 J1=1,3
2230          DO 36 K1=J1,3
2240          JK=JK+1
2250          IF(J1.EQ.J.AND.K1.EQ.K)GO TO 40
2260    36    CONTINUE
2270    35    CONTINUE
2280    40    L=L+1
2290          E(L)=A(L)*B(1)+A(IJ)*B(K+1)+A(IK)*B(J+1)+A(I+1)*B(JK)
2292          E(L)=E(L)+A(JK)*B(I+1)+A(J+1)*B(IK)+A(K+1)*B(IJ)
2295          E(L)=E(L)+A(1)*B(L)
2297    20    CONTINUE
2300    19    CONTINUE
2310    18    CONTINUE
2320          RETURN
2330          END
2340          SUBROUTINUE ADD(D,E,G)
2350          DIMENSION D(20),E(20),G(20)
2360          DO 50 I=1,20
2370    50    G(I)=D(I)+E(I)
2380          RETURN
2390          END
2400          SUBROUTINE SR(U,S)
2410          DIMENSION U(20),S(20),F(4)
2420          F(1)=U(1)©.5
```

```
2430          F(2)=.5*U(1)©(-.5)
2440          F(3)=-(.25)*U(1)©(-1.5)
2450          F(4)=(3./8.)*U(1)©(-2.5)
2460          CALL DER(F,U,S)
2470          RETURN
2480          END
2490          SUBROUTINE DER(F,U,S)
2500          DIMENSION F(4),U(20),S(20)
2510          S(1)=F(1)
2520          DO 55 I=1,3
2530    55    S(I+1)=F(2)*U(I+1)
2540          L=4
2550          DO 56 I=1,3
2560          DO 57 J=I,3
2570          L=L+1
2580    57    S(L)=F(3)*U(J+1)*U(I+1)+F(2)*U(L)
2590    56    CONTINUE
2600          L=10
2610          IJ=4
2620          DO 58 I=1,3
2630          DO 59 J=I,3
2640          IJ=IJ=+1
2650          DO 60 K=J,3
2660          IK=4
2670          DO 65 I1=1,3
2680          DO 66 K1=I1,3
2690          IK=IK+1
2700          IF(I1.EQ.I.AND.K1.EQ.K)GO TO 70
2710    66    CONTINUE
2720    65    CONTINUE
2730    70    JK=4
2740          DO 75 J1=1,3
2750          DO 76 K1=J1,3
2760          JK=JK+1
2770          IF(J1.EQ.J.AND.K1.EQ.K)GO TO 80
2780    76    CONTINUE
2790    75    CONTINUE
2800    80    L=L+1
2900          S(L)=F(4)*U(K+1)*U(J+1)*U(I+1)+F(3)*U(I+1)*U(JK)
2905    60    S(L)=S(L)+F(3)*U(J+1)*U(IK)+F(3)*U(K+1)*U(IJ)+F(2)*U(L)
2910    59    CONTINUE
2920    58    CONTINUE
2930          RETURN
2940          END
2950          SUBROUTINE DIV(U,B,F1)
2960          DIMENSION U(20),B(20),R(20),F1(20),F(4)
2970          F(1)=U(1)©(-1.)
2980          F(2)=-U(1)©(-2.)
2990          F(3)=2.*U(1)©(-3.)
3000          F(4)=-6.*U(1)©(-4.)
3010          CALL DER(F,U,R)
3015          CALL MULT(R,B,F1)
3020          RETURN
3030          END
3200          SUBROUTINE INTEG(X1,X2,X3,CON,X1D,A,B,C,F1,Z,DZ)
3210          DIMENSION A(20),B(20),C(20),F1(20),Z(4),DZ(4)
```

```
3212          DIMENSION M(10),Y(4)
3213          DIMENSION AH(4),AI(4),AJ(4),AK(4)
3215          H=1./100.
3220          C1=1.0
3230          H2=H/2.0
3240          L=INT(X1D/H+.1*H)
3250          K=10
3260          I1=1
3265          XA=X1
3270          TYPE 88
3280    88    FORMAT(40H    X         Z1         Z2         Z3         Z4)
3290          TYPE*,X1,Z(1),Z(2),Z(3),Z(4)
3300          DO 90 I=1,K
3310          M(I)=INT((X1D/K*I+.1*H)/H)
3315    90    CONTINUE
3320          DO 92 N=1,L
3330          X1=XA+(N-1)*H
3340          X=X1
3350          CALL MMULT(Z,C1,Y)
3360          CALL FORCE(X1,X2,X3,CON,A,B,C,F1,Z,DZ)
3370          CALL MMULT(DZ,H2,AH)
3380          CALL MADD(Y,AH,Z)
3390          X=X1+H2
3400          CALL FORCE(X1,X2,X3,CON,A,B,C,F1,Z,DZ)
3410          CALL MMULT(DZ,H2,AI)
3420          CALL MADD(Y,AI,Z)
3430          CALL FORCE(X1,X2,X3,CON,A,B,C,F1,Z,DZ)
3440          CALL MMULT(DZ,H,AJ)
3450          CALL MADD(Y,AJ,Z)
3460          X=X1+H
3470          CALL FORCE(X1,X2,X3,CON,A,B,C,F1,Z,DZ)
3475          CALL MMULT(DZ,H,AK)
3480          CALL MMUAD(AH,AI,AJ,AK,Y,Z)
3490          M1=M(I1)
3500          IF(N-M1)92,95,92
3505    95    TYPE*,X,Z(1),Z(2),Z(3),Z(4)
3510          I1=I1+1
3515    92    CONTINUE
3520          RETURN
3530          END
3550          SUBROUTINE MMULT(Z,C1,Y)
3560          DIMENSION Z(4),Y(4)
3570          DO 95 I=1,4
3580    95    Y(I)=C1*Z(I)
3590          RETURN
3600          END
3610          SUBROUTINE MADD(Y,AH,Z)
3620          DIMENSION Y(4),AH(4),Z(4)
3630          DO 100 I=1,4
3640   100    Z(I)=Y(I)+AH(I)
3650          RETURN
3660          END
3670          SUBROUTINE MMUAD(AH,AI,AJ,AK,Y,Z)
3680          DIMENSION AH(4),AI(4),AJ(4),AK(4),Y(4),Z(4)
3690          DO 105 I=1,4
3700   105    Z(I)=Y(I)+(2.*AH(I)+4.*AI(I)+2.*AJ(I)+AK(I))/6.
```

```
3710          RETURN
3720          END
3730          SUBROUTINE FORCE(X1,X2,X3,CON,A,B,C,F1,Z,DZ)
3740          DIMENSION A(20),B(20),C(20),F1(20),Z(4),DZ(4)
3750          X2=Z(1)
3760          X3=Z(2)
3770          CALL INPUT(X1,X2,X3,CON,A,B,C,F1)
3780          DZ(1)=Z(2)
3790          DZ(2)=(F1(3)-F1(7)-F1(9)*Z(2))/F1(10)
3800          DZ(3)=Z(4)
3810          DZ(4)=F1(8)*Z(3)+F1(9)*Z(4)-F1(15)*Z(3)-F1(16)*Z(4)
3812          DZ(4)=DZ(4)-F1(18)*Z(2)*Z(3)-F1(19)*Z(2)*Z(4)-F1(9)*Z(4)
3815          ADZ4=F1(3)-F1(7)-F1(9)*Z(2)
3820          DZ(4)=(F1(10)*DZ(4)-(ADZ4*(F1(19)*Z(3)+F1(20)*Z(4))))
3825          DZ(4)=DZ(4)/F1(10)©2.
3830          RETURN
3840          END
```

3.5.2. Second-Order Systems Using the Newton–Raphson Method

Method of Solution: Newton–Raphson method using Pontryagin's maximum principle, Section 3.2.1.2.

Table 3.31. Definitions of the FORTRAN Variables and Vectors for the Given Example of a Second-Order Optimal Control Problem

	Variable	FORTRAN variable or vector	Number of components	Program line number
Initial conditions	$\mathbf{x}(0)$	CI	12	135–160
Integration limits	0	X1I	1	175
	T	X1L	1	190
Integrand	$F = x_1^2 + y$	F1	18	900
Constraints	$g_1 = x_2$	G1	18	910
	$g_2 = -b_1 x_1 - b_2 x_1^3 + y$	G2	18	940
Integration grid interval size	Δt	H	1	3215
State and costate variables[a]	$\mathbf{x}$	Z	12	3480
Derivatives to be integrated each iteration	$\dot{\mathbf{x}}$	DZ	12	3780–3895

[a] Where $y = -\frac{1}{2}p_2$ must be expressed as a function of the state and costate variables. The vector components of $\mathbf{x} = (x_1, x_2, p_1, p_2, x_{1c_1}, x_{2c_1}, p_{1c_1}, p_{2c_1}, x_{1c_2}, x_{2c_2}, p_{1c_2}, p_{2c_2})^T$ are defined by equations (3.72)–(3.75) and (3.84)–(3.87). Additional definitions of the variables and vectors are given in Tables 3.5 and 3.6.

Description of example and numerical results: Section 3.4.1.3.

Definitions of the FORTRAN variables and vectors for the given example: Table 3.31.

FORTRAN program: Listing 2.

FORTRAN Program Listing 2

Second-Order Optimal Control Problem, Newton Raphson Method

```
100        DIMENSION A1(18),A2(18),A3(18),A4(18),A5(18)
105        DIMENSION F1(18),G1(18),G2(18)
110        DIMENSION CI(12),Z(12),DZ(12),X(5)
130        DO 5 I=1,12
135    5   CI(I)=0.0
140        CI(1)=1.0
150        CI(7)=1.0
160        CI(12)=1.0
175        X1I=0.0
190        X1L=1.0
210        X1D=X1L-X1I
220        CON=0.25
300        DO 150 N1=1,6
310        CALL MMULT(CI,1.,Z)
320        X(1)=X1I
330        CALL INTEG(X,CON,XID,A1,A2,A3,A4,A5,F1,G1,G2,Z,DZ)
332        DO 8 I=1,4
335    8   X(I+1)=Z(I)
340        CALL INPUT(X,CON,A1,A2,A3,A4,A5,F1,G1,G2)
345        D=Z(7)*Z(12)-Z(11)*Z(8)
350        CI(3)=CI(3)-(Z(3)*Z(12)-Z(4)*Z(11))/D
355  150   CI(4)=CI(4)-(Z(4)*Z(7)-Z(3)*Z(8))/D
360        END
850        SUBROUTINE INPUT(X,CON,A1,A2,A3,A4,A5,F1,G1,G2)
855        DIMENSION X(5),A1(18),A2(18),A3(18),A4(18),A5(18)
860        DIMENSION D(18),F1(18),G1(18),G2(18)
865        DIMENSION E1(18),E2(18),E3(18)
867        DIMENSION E4(18),E5(18),E6(18),E7(18),E8(18),E9(18)
870        CALL LIN(X,A1,A2,A3,A4,A5)
880        CALL MULT(A2,A2,E1)
885        CALL CONST(CON,D)
890        CALL MULT(A5,A5,E2)
895        CALL MULT(D,E2,E3)
900        CALL ADD(E1,E3,F1)
905        CALL CONST(1.,D)
910        CALL MULT(D,A3,G1)
915        CALL CONST(-.5,D)
920        CALL MULT(D,A5,E4)
922        B1=-1.0
924        B2=-4.0
926        CALL CONST(B1,D)
928        CALL MULT(D,A2,E5)
```

```
 930          CALL ADD(E4,E5,E6)
 932          CALL CONST(B2,D)
 934          CALL MULT(A2,A2,E7)
 936          CALL MULT(E7,A2,E8)
 938          CALL MULT(D,E8,E9)
 940          CALL ADD(E6,E9,G2)
 942          RETURN
 945          END
 950          SUBROUTINE LIN(X,A1,A2,A3,A4,A5)
 955          DIMENSION X(5),A1(18),A2(18),A3(18),A4(18),A5(18)
 960          DO 10 I=1,18
 965          A1(I)=0.0
 970          A2(I)=0.0
 975          A3(I)=0.0
 980          A4(I)=0.0
 985    10    A5(I)=0.0
 990          A1(1)=X(1)
 995          A1(2)=1.0
1000          A2(1)=X(2)
1005          A2(3)=1.0
1010          A3(1)=X(3)
1015          A3(4)=1.0
1020          A4(1)=X(4)
1025          A4(5)=1.0
1030          A5(1)=X(5)
1035          A5(6)=1.0
1070          RETURN
1080          END
1090          SUBROUTINE CONST(CON,D)
1100          DIMENSION D(18)
1110          DO 12 I=1,18
1120    12    D(I)=0.0
1130          D(1)=CON
1140          RETURN
1150          END
1160          SUBROUTINE MULT(A,B,E)
1170          DIMENSION A(18),B(18),E(18)
1180          E(1)=A(1)*B(1)
1190          DO 15 I=1,5
2000    15    E(I+1)=A(I+1)*B(1)+A(1)*B(I+1)
2010          L=6
2020          DO 16 I=1,3
2030          DO 17 J=I,5
2040          L=L+1
2050    17    E(L)=A(L)*B(1)+A(I+1)*B(J+1)+A(J+1)*B(I+1)+A(1)*B(L)
2070    16    CONTINUE
2320          RETURN
2330          END
2340          SUBROUTINE ADD(D,E,G)
2350          DIMENSION D(18),E(18),G(18)
2360          DO 50 I=1,18
2370    50    G(I)=D(I)+E(I)
2380          RETURN
2390          END
2400          SUBROUTINE SR(U,S)
2410          DIMENSION U(18),S(18),F(3)
```

```
2420          F(1)=U(1)©.5
2430          F(2)=.5*U(1)©(-.5)
2440          F(3)=-(.25)*U(1)©(-1.5)
2460          CALL DER(F,U,S)
2470          RETURN
2480          END
2490          SUBROUTINE DER(F,U,S)
2500          DIMENSION F(3),U(18),S(18)
2510          S(1)=F(1)
2520          DO 55 I=1,5
2530    55    S(I+1)=F(2)*U(I+1)
2540          L=6
2550          DO 56 I=1,3
2560          DO 57 J=I,5
2570          L=L+1
2580    57    S(L)=F(3)*U(J+1)*U(I+1)+F(2)*U(L)
2590    56    CONTINUE
2930          RETURN
2940          END
2950          SUBROUTINE DIV(U,B,F1)
2960          DIMENSION U(18),B(18),R(18),F1(18),F(3)
2970          F(1)=U(1)©(-1.)
2980          F(2)=-U(1)©(-2.)
2990          F(3)=2.*U(1)©(-3.)
3010          CALL DER(F,U,R)
3015          CALL MULT(R,B,F1)
3020          RETURN
3030          END
3200          SUBROUTINE INTEG(X,CON,X1D,A1,A2,A3,A4,A5,F1,G1,G2,Z,DZ)
3210          DIMENSION A1(18),A2(18),A3(18),A4(18),A5(18)
3212          DIMENSION Z(12),DZ(12),M(10),Y(12)
3213          DIMENSION AH(12),AI(12),AJ(12),AK(12)
3214          DIMENSION X(5),F1(18),G1(18),G2(18)
3215          H=1./100.
3220          C1=1.0
3230          H2=H/2.0
3240          L=INT(X1D/H+.1*H)
3250          K=10
3260          I1=1
3265          XA=X(1)
3270          TYPE 88
3280    88    FORMAT(38H      X                Z1                Z2)
3285          Z5=-.5*Z(4)
3290          TYPE*,X(1),Z(1),Z(2),Z5
3300          DO 90 I=1,K
3310          M(I)=INT((X1D/K*I+.1*H)/H)
3315    90    CONTINUE
3320          DO 92 N=1,L
3330          XS=XA+(N-1)*H
3340          X(1)=XS
3350          CALL MMULT(Z,C1,Y)
3360          CALL FORCE(X,CON,A1,A2,A3,A4,A5,F1,G1,G2,Z,DZ)
3370          CALL MMULT(DZ,H2,AH)
3380          CALL MADD(Y,AH,Z)
3390          X(1)=XS+H2
3400          CALL FORCE(X,CON,A1,A2,A3,A4,A5,F1,G1,G2,Z,DZ)
```

```
3410          CALL MMULT(DZ,H2,AI)
3420          CALL MADD(Y,AI,Z)
3430          CALL FORCE(X,CON,A1,A2,A3,A4,A5,F1,G1,G2,Z,DZ)
3440          CALL MMULT(DZ,H,AJ)
3450          CALL MADD(Y,AJ,Z)
3460          X(1)=XS+H
3470          CALL FORCE(X,CON,A1,A2,A3,A4,A5,F1,G1,G2,Z,DZ)
3475          CALL MMULT(DZ,H,AK)
3480          CALL MMUAD(AH,AI,AJ,AK,Y,Z)
3490          M1=M(I1)
3500          IF(N-M1)92,95,92
3502    95    Z5=-.5*Z(4)
3505          TYPE*,X(1),Z(1),Z(2),Z5
3510          I1=I1+1
3515    92    CONTINUE
3520          RETURN
3530          END
3550          SUBROUTINE MMULT(Z,C1,Y)
3560          DIMENSION Z(12),Y(12)
3570          DO 95 I=1,12
3580    95    Y(I)=C1*Z(I)
3590          RETURN
3600          END
3610          SUBROUTINE MADD(Y,AH,Z)
3620          DIMENSION Y(12),AH(12),Z(12)
3630          DO 100 I=1,12
3640   100    Z(I)=Y(I)+AH(I)
3650          RETURN
3660          END
3670          SUBROUTINE MMUAD(AH,AI,AJ,AK,Y,Z)
3680          DIMENSION AH(12),AI(12),AJ(12),AK(12),Y(12),Z(12)
3690          DO 105 I=1,12
3700   105    Z(I)=Y(I)+(2.*AH(I)+4.*AI(I)+2.*AJ(I)+AK(I))/6.
3710          RETURN
3720          END
3730          SUBROUTINE FORCE(X,CON,A1,A2,A3,A4,A5,F1,G1,G2,Z,DZ)
3740          DIMENSION A1(18),A2(18),A3(18),A4(18),A5(18)
3745          DIMENSION X(5),F1(18),G1(18),G2(18),Z(12),DZ(12)
3750          DO 110 I=1,4
3755          X(I+1)=Z(I)
3760   110    CONTINUE
3770          CALL INPUT(X,CON,A1,A2,A3,A4,A5,F1,G1,G2)
3780          DZ(1)=G1(1)
3785          DZ(2)=G2(1)
3790          DZ(3)=-(F1(3)+Z(3)*G1(3)+Z(4)*G2(3))
3800          DZ(4)=-(F1(4)+Z(3)*G1(4)+Z(4)*G2(4))
3805          DO 115 I=5,12
3810   115    DZ(I)=0.0
3815          DZ(7)=-Z(7)*G1(3)-Z(8)*G2(3)
3820          DZ(11)=-Z(11)*G1(3)-Z(12)*G2(3)
3825          DO 120 I=1,4
3830          DZ(5)=DZ(5)+G1(I+2)*Z(I+4)
3835          DZ(6)=DZ(6)+G2(I+2)*Z(I+4)
3840          D7=F1(I+11)+Z(3)*G1(I+11)+Z(4)*G2(I+11)
3845          DZ(7)=DZ(7)-D7*Z(I+4)
3850          DZ(9)=DZ(9)+G1(I+2)*Z(I+8)
```

```
3855           DZ(10)=DZ(10)+G2(I+2)*Z(I+8)
3860     120   DZ(11)=DZ(11)-D7*Z(I+8)
3865           D8=F1(13)+Z(3)*G1(13)+Z(4)*G2(13)
3870           DZ(8)=-Z(7)*G1(4)-Z(8)*G2(4)-D8*Z(5)
3875           DZ(12)=-Z(11)*G1(4)-Z(12)*G2(4)-D8*Z(9)
3880           DO 125 I=1,3
3885           D9=F1(I+15)+Z(3)*G1(I+15)+Z(4)*G2(I+15)
3890           DZ(8)=DZ(8)-D9*Z(I+5)
3895     125   DZ(12)=DZ(12)-D9*Z(I+9)
3900           RETURN
3905           END
```

3.5.3. *N*th-Order Systems Using the Gradient Method

Method of Solution: Gradient method with Newton–Raphson recurrence relation, Section 3.2.2.2.

Description of example and numerical results: Section 3.4.2.3.

Definitions of the FORTRAN variables and vectors for the given example: Table 3.32.

FORTRAN program: Listing 3.

The program is dimensioned to handle any order system from 1 to 4, where ID = system order in line 75. The program can easily be dimensioned to handle an *n*th-order system. To increase the dimensions of the program from fourth to fifth order systems for example, dimension as follows.

Main Program

```
55    Dimension X(12),HA(14),X1(233,5),A(12,14)
60    Dimension CI(5),Z(5),DZ(5),X0(5),P0(5)
65    Dimension AH(5),AI(5),AJ(5),AK(5),Y(5)
75    ID=5
```

Subroutine INPUT

```
790   Dimension A(12,14),T(14),Y(14)
795   Dimension X1(14),X2(14),X3(14),X4(14),X5(14)
800   Dimension P1(14),P2(14),P3(14),P4(14),P5(14)
805   Dimension X(IV),HA(LV)F(14),D(14)
```

```
810     Dimension E1(14),E2(14),E3(14),E4(14),E5(14),E6(14)
812     Dimension E7(14),E8(14),D1(14),D2(14)
830     Go to(1,2,3,4,5)ID
840 5 X5(I)=A(6,I)
845     P5(I)=A(6+ID,I)
```

Subroutine LIN

```
995     Dimension X(IV),A(12,14)
```

Table 3.32. Definitions of the FORTRAN Variables and Vectors for the Given Example of a Third-Order Optimal Control Problem

	Variable	FORTRAN variable or vector	Number of components	Program line number
System order	n	ID	1	75
Initial conditions	$\mathbf{x}(0)$	CI	3	130–155
Integration grid interval size	Δt	H	1	170
Integration limits	0	X1I	1	180
	T	X1L	1	190
Terminal conditions	$\mathbf{p}(T)$	PO	3	385–387
Hamiltonian	$H = p_1 x_3 \cos y + p_2 x_3 \sin y + p_3 \sin y$	HA	10	945
Initial approximation of the control function	$y(t)$	U	171	240
State variables[a]	$\mathbf{x}$	Z	3	3480
Costate variables	$\mathbf{p}$	Z	3	3480
Derivatives to be integrated each iteration:				
Forward integration	$\dot{\mathbf{x}}$	DZ	3	3825
Backward integration	$\dot{\mathbf{p}}$	DZ	3	3825

[a] Where $\mathbf{x} = (x_1, x_2, x_3)^T$ and $\mathbf{p} = (p_1, p_2, p_3)^T$. The control function, U, is dimensioned to store up to 233 grid values. Additional definitions of the variables and vectors are given in Tables 3.11 and 3.12.

FORTRAN Program Listing 3
Nth-Order Optimal Control Problem, Gradient Method

```
50            DIMENSION U(233),U1(233),UT(233)
55            DIMENSION X(10),HA(12),X1(233,4),A(10,12)
60            DIMENSION CI(4),Z(4),DZ(4),XO(4),PO(4)
65            DIMENSION AH(4),AI(4),AJ(4),AK(4),Y(4)
70            COMMON L,XID,ID,L1,LV,IV,A
75            ID=3
80            LV=4+2*ID
85            IV=LV-2
90            DO 335 I=1,IV
95            DO 340 K=1,LV
100    340     A(I,K)=0.0
105    335     CONTINUE
110            DO 345 I=1,IV
120    345     A(I,I+1)=1.0
130            CI(1)=0.
140            CI(2)=0.0
150            CI(3)=0.07195
155            CI(4)=0.0
170            H=1./100.
180            X1I=0.0
190            X1L=1.7
210            X1D=X1L-X1I
220            L=INT(X1D/H+.1*H)
225            L1=L+1
230            DO 140 I=1,173
240    140     U(I)=0.5235988
250            JS=1
260            DO 355 I=1,ID
265    355     X1(1,I)=CI(I)
270            DO 150 N1=1,10
280            DO 365 I=1,ID
285    365     XO(I)=CI(I)
305            CALL MMULT(XO,1.,Z)
315            KP=5
320            X(1)=X1I
330            IN=1
335            DT=H
340            CALL INTEG(X,HA,X1,U,U1,Z,DZ,IN,DT,KP,JS,AH,AI,AJ,AK,Y)
350            IN=2
360            DT=-H
370            X(1)=X1L
380            DO 360 I=1,ID
385    360     PO(I)=0.0
387            PO(1)=-1.
390            CALL MMULT(PO,1.,Z)
400            CALL INTEG(X,HA,X1,U,U1,Z,DZ,IN,DT,KP,JS,AH,AI,AJ,AK,Y)
410            DO 350 I=1,ID
415            I1=I+1
420            I2=I+ID+1
425            X(I1)=CI(I)
430    350     X(I2)=Z(I)
435            X(IV)=U(1)
```

```
440            CALL INPUT(X,HA)
450            U1(1)=-HA(IV+1)/HA(IV+2)
470            DO 160 I=1,L1
480   160      U(I)=U(I)+U1(I)
710   150      CONTINUE
720            END
780            SUBROUTINE INPUT(X,HA)
785            COMMON L,X1D,ID,L1,LV,IV,A
790            DIMENSION A(10,12),T(12),Y(12)
795            DIMENSION X1(12),X2(12),X3(12),X4(12)
800            DIMENSION P1(12),P2(12),P3(12),P4(12)
805            DIMENSION X(IV),HA(LV),F(12),D(12)
810            DIMENSION E1(12),E2(12),E3(12),E4(12),E5(12),E6(12)
812            DIMENSION E7(12),E8(12),D1(12),D2(12)
815            CALL LIN(X)
820            DO 325 I=1,LV
825            T(I)=A(1,I)
830            GO TO(1,2,3,4)ID
850     4      X4(I)=A(5,I)
855            P4(I)=A(5+ID,I)
860     3      X3(I)=A(4,I)
865            P3(I)=A(4+ID,I)
870     2      X2(I)=A(3,I)
875            P2(I)=A(3+ID,I)
880     1      X1(I)=A(2,I)
885            P1(I)=A(2+ID,I)
890   325      Y(I)=A(IV,I)
900            CALL LIN(X)
905            CALL SCOS(Y,E1)
910            CALL MULT(X3,E1,E2)
915            CALL MULT(P1,E2,E3)
920            CALL SSIN(Y,E1)
925            CALL MULT(X3,E1,E2)
930            CALL MULT(P2,E2,E4)
935            CALL ADD(E3,E4,E5)
940            CALL MULT(P3,E1,E2)
945            CALL ADD(E5,E2,HA)
950            RETURN
955            END
985            SUBROUTINE LIN(X)
990            COMMON L,X1D,ID,L1,LV,IV,A
995            DIMENSION X(IV),A(10,12)
1000            DO 330 I=1,IV
1005  330       A(I,1)=X(I)
1010            RETURN
1015            END
1090            SUBROUTINE CONST(CON,D)
1095            COMMON L,X1D,ID,L1,LV,IV
1100            DIMENSION D(LV)
1110            DO 15 I=1,LV
1120   15       D(I)=0.0
1130            D(1)=CON
1140            RETURN
1150            END
1160            SUBROUTINE MULT(A,B,E)
1165            COMMON L,X1D,ID,L1,LV,IV
```

```
1170          DIMENSION A(LV),B(LV),E(LV)
1180          E(1)=A(1)*B(1)
1190          DO 15 I=1,IV
2000   15     E(I+1)=A(I+1)*B(1)+A(1)*B(I+1)
2010          E(LV)=A(LV)*B(1)+2.*A(LV-1)*B(LV-1)+A(1)*B(LV)
2020          RETURN
2030          END
2040          SUBROUTINE SCOS(U,S)
2050          COMMON L,X1D,ID,L1,LV,IV
2060          DIMENSION U(LV),S(LV),F(3)
2070          F(1)=COS(U(1))
2080          F(2)=-SIN(U(1))
2090          F(3)=-COS(U(1))
2100          CALL DER(F,U,S)
2110          RETURN
2120          END
2130          SUBROUTINE SSIN(U,S)
2140          COMMON L,X1D,ID,L1,LV,IV
2150          DIMENSION U(LV),S(LV),F(3)
2160          F(1)=SIN(U(1))
2170          F(2)=COS(U(1))
2180          F(3)=-SIN(U(1))
2190          CALL DER(F,U,S)
2200          RETURN
2220          END
2260          SUBROUTINE AEXP(U,S)
2270          DIMENSION U(8),S(8),F(3)
2280          F(1)=EXP(U(1))
2290          F(2)=F(1)
2300          F(3)=F(1)
2310          CALL DER(F,U,S)
2320          RETURN
2330          END
2340          SUBROUTINE ADD(D,E,G)
2345          COMMON L,X1D,ID,L1,LV,IV
2350          DIMENSION D(LV),E(LV),G(LV)
2360          DO 50 I=1,LV
2370   50     G(I)=D(I)+E(I)
2380          RETURN
2390          END
2400          SUBROUTINE SR(U,S)
2410          DIMENSION U(20),S(20),F(4)
2420          F(1)=U(1)^.5
2430          F(2)=.5*U(1)^(-.5)
2440          F(3)=-(.25)*U(1)^(-1.5)
2450          F(4)=(3./8.)*U(1)^(-2.5)
2460          CALL DER(F,U,S)
2470          RETURN
2480          END
2490          SUBROUTINE DER(F,U,S)
2495          COMMON L,X1D,ID,L1,LV,IV
2500          DIMENSION U(LV),S(LV),F(3)
2510          S(1)=F(1)
2520          DO 55 I=1,IV
2530   55     S(I+1)=F(2)*U(I+1)
2540          S(LV)=F(3)*U(LV-1)^2+F(2)*U(LV)
```

```
2930          RETURN
2940          END
2950          SUBROUTINE DIV(U,B,F1)
2960          DIMENSION U(20),B(20),R(20),F1(20),F(4)
2970          F(1)=U(1)^(-1.)
2980          F(2)=-U(1)^(-2.)
2990          F(3)=2.*U(1)^(-3.)
3000          F(4)=-6.*U(1)^(-4.)
3010          CALL DER(F,U,R)
3015          CALL MULT(R,B,F1)
3020          RETURN
3030          END
3180          SUBROUTINE INTEG(X,HA,X1,U,U1,Z,DZ,IN,H,K,J,AH,AI,AJ,AK,Y)
3185          COMMON L,X1D,ID,L1,LV,IV
3190          DIMENSION U(L1),U1(L1),HA(LV),X(IV)
3195          DIMENSION Z(ID),DZ(ID),M(10)
3200          DIMENSION AH(ID),AI(ID),AJ(ID),AK(ID),Y(ID)
3205          DIMENSION X1(L1,ID)
3220          C1=1.0
3230          H2=H/2.0
3235          H1=ABS(H)
3260          I1=1
3265          XA=X(1)
3270          TYPE 88
3280    88    FORMAT(38H      X                  Z1                Z2)
3290          TYPE*,X(1),Z(1),Z(2),U(1)
3300          DO 90 I=1,K
3310          M(I)=INT((X1D/K*I+.1*H1)/H1)
3315    90    CONTINUE
3320          DO 98 N=1,L
3325          NL=N
3330          XS=XA+(N-1)*H
3340          X(1)=XS
3350          CALL MMULT(Z,C1,Y)
3355          DX=0.0
3360          CALL FORCE(X,HA,X1,U,U1,Z,DZ,IN,NL,DX)
3362          IF(IN.EQ.1)GO TO 97
3365          U1(L+2-N)=-HA(IV+1)/HA(IV+2)
3370    97    CALL MMULT(DZ,H2,AH)
3380          CALL MADD(Y,AH,Z)
3390          X(1)=XS+H2
3395          DX=0.5
3400          CALL FORCE(X,HA,X1,U,U1,Z,DZ,IN,NL,DX)
3410          CALL MMULT(DZ,H2,AI)
3420          CALL MADD(Y,AI,Z)
3430          CALL FORCE(X,HA,X1,U,U1,Z,DZ,IN,NL,DX)
3440          CALL MMULT(DZ,H,AJ)
3450          CALL MADD(Y,AJ,Z)
3460          X(1)=XS+H
3465          DX=1.0
3470          CALL FORCE(X,HA,X1,U,U1,Z,DZ,IN,NL,DX)
3475          CALL MMULT(DZ,H,AK)
3480          CALL MMUAD(AH,AI,AJ,AK,Y,Z)
3490          M1=M(I1)
3500          IF(N-M1)92,95,92
3505    95    TYPE*,X(1),Z(1),Z(2),U(N+1)
```

```
3510          I1=I1+1
3515    92    IF(IN.EQ.2)GO TO 98
3520          DO 320 I=1,ID
3525   320    X1(N+1,I)=Z(I)
3530    98    CONTINUE
3535          RETURN
3540          END
3550          SUBROUTINE MMULT(Z,C1,Y)
3555          COMMON L,XID,ID,L1,LV,IV
3560          DIMENSION Z(ID),Y(ID)
3570          DO 94 I=1,ID
3580    94    Y(I)=C1*Z(I)
3590          RETURN
3600          END
3610          SUBROUTINE MADD(Y,AH,Z)
3615          COMMON L,XID,ID,L1,LV,IV
3620          DIMENSION Y(ID),AH(ID),Z(ID)
3630          DO 100 I=1,ID
3640   100    Z(I)=Y(I)+AH(I)
3650          RETURN
3660          END
3670          SUBROUTINE MMUAD(AH,AI,AJ,AK,Y,Z)
3675          COMMON L,X1D,ID,L1,LV,IV
3680          DIMENSION AH(ID),AI(ID),AJ(ID),AK(ID)
3685          DIMENSION Y(ID),Z(ID)
3690          DO 105 I=1,ID
3700   105    Z(I)=Y(I)+(2.*AH(I)+4.*AI(I)+2.*AJ(I)+AK(I))/6.
3710          RETURN
3720          END
3730          SUBROUTINE FORCE(X,HA,X1,U,U1,Z,DZ,IN,N,DX)
3740          COMMON L,X1D,ID,L1,LV,IV
3745          DIMENSION X1(L1,ID),U(L1),U1(L1)
3750          DIMENSION Z(ID),DZ(ID),HA(LV),X(IV)
3770          IF(IN.EQ.2)GO TO 200
3775          DO 300 I=1,ID
3780          I1=I+1
3785          I2=I+ID+1
3790          X(I1)=Z(I)
3795   300    X(I2)=0.0
3800          X(IV)=U(N)+(U(N+1)-U(N))*DX
3810          CALL INPUT(X,HA)
3815          DO 305 I=1,ID
3820          I1=I+ID+2
3825   305    DZ(I)=HA(I1)
3830          GO TO 210
3840   200    DO 310 I=1,ID
3845          I1=I+1
3850          I2=I+ID+1
3855          X(I1)=X1(L+2-N,I)+(X1(L+1-N,I)-X1(L+2-N,I))*DX
3860   310    X(I2)=Z(I)
3865          X(IV)=U(L+2-N)+(U(L+1-N)-U(L+2-N))*DX
3870          CALL INPUT(X,HA)
3875          DO 315 I=1,ID
3880          I1=I+2
3885   315    DZ(I)=-HA(I1)
3890   210    RETURN
3900          END
```

Exercises

1. The shape of a rigid wire with fixed ends, $x(0) = 0$, $x(1) = 1$, is to be determined such that a frictionless bead slides down the wire from 0 to 1 under the force of gravity, g, in minimum time.

   (a) Express the problem in terms of the simplest problem in the calculus of variations.

   (b) Obtain the numerical solution using the automatic derivative evaluation program.

   Answer: For a discussion of this problem, see p. 361, Reference 5.

2. Find the trajectory, $x(t)$, that minimizes the cost functional

$$J = \frac{1}{2} \int_0^{0.1} (x^2 + y^2)\, dt$$

   subject to the differential constraints and initial condition

$$\dot{x}(t) = -x^2 + y, \qquad x(0) = 1.$$

   Use the automatic derivative evaluation program.

   Answer: p. 85, Reference 1.

3. A vehicle in space, considered as a particle of mass m, is acted upon by a thrust of magnitude ma. The control variable is the thrust direction angle $y(t)$. The planar equations of motion for constant thrust acceleration, a, are

$$\begin{aligned} \dot{x}_1 &= x_3, & x_1(0) &= 0, \\ \dot{x}_2 &= x_4, & x_2(0) &= h, \\ \dot{x}_3 &= a \cos y, & x_3(0) &= 0, \\ \dot{x}_4 &= a \sin y, & x_4(0) &= 0 \end{aligned}$$

   where x_1, x_2 are the inertial coordinates and x_3, x_4 are the velocity components. Find the trajectory such that the (negative) velocity $x_3(T)$ is maximized and the offset $x_2(t)$ is minimized. The cost functional is

$$J = x_3(T) + \int_0^T S x_2(t)\, dt$$

   Assume $a = 32$ ft/sec^2 and $h = 4.262595$ ft.

   Answer: p. 349, Reference 18. See also p. 59, Reference 3.

4

System Identification

Over the last 30 years there has been great interest in system identification, especially insofar as adaptive control systems are concerned. In such systems the controller has to make inferences concerning unknown aspects of a process and then make decisions based on these inferences.

Often the dynamics of the system are described by nonlinear ordinary differential equations containing unknown parameters. If noisy observations are available on the state of the system at various times, the task is to estimate the parameters in the model to give the best fit of the model to the data. Here "best fit" is frequently interpreted in the least-squares sense.

Various approaches to this problem are known, with quasilinearization (References 1–3) being a prominent choice. It is a successive approximation scheme, and long experience has shown that rapid convergence, even with poor initial estimates of the parameters, is the rule rather than the exception (References 4 and 9).

In this chapter we present a general FORTRAN program for fitting systems of nonlinear ordinary differential equations to observations. Rather than using the basic quasilinearization scheme, we employ a variation which involves the simultaneous calculation of approximations method (SCAM). We do this to cut down on memory requirements, for the storage of functions is eliminated. Additionally, there is an improvement in accuracy, since in the Runge–Kutta scheme employed, required values of right-hand sides at midpoints of intervals need not be approximated by some interpolation formula.

The second improvement is the incorporation of the numerical scheme for the fast and efficient evaluation of derivatives (FEED). This is important because we have to evaluate not only the right-hand sides of the ordinary differential equations but also their derivatives with respect to the dependent variables and the unknown parameters. With ten dependent variables and five unknown parameters, for example, a total of 150 derivatives would have to be evaluated, an error-prone process. The current program eliminates this roadblock, as will be explained later.

In Section 4.1 the description of quasilinearization combined with SCAM for a system of nonlinear ordinary differential equations is given. The FEED procedure used for evaluating the numerical values of the required derivatives is described in Section 4.2. Section 4.3 is a detailed explanation of the computer program that handles the estimation problem for a system of up to ten ordinary differential equations and five parameters. The numerical results for a system of two ordinary differential equations are given in Section 4.4, and the FORTRAN program listing is in Section 4.5.

4.1. Quasilinearization

Let us consider the system of ordinary differential equations,

$$\dot{x}_i = f^i(x_1, x_2, \ldots, x_n; \alpha_1, \alpha_2, \ldots, \alpha_m), \qquad i = 1, 2, 3, \ldots, n, \tag{4.1}$$

subject to the initial conditions

$$x_i(0) = \gamma_i, \qquad i = 1, 2, 3, \ldots, n. \tag{4.2}$$

In this system of equations we assume that $\alpha_1, \alpha_2, \ldots, \alpha_m$ are unknown parameters to be estimated, and $\gamma_1, \gamma_2, \ldots, \gamma_n$ are known initial conditions (γ_i = unknown is a trivial extension of the problem considered).

Let us assume that at T times, observations are made on the kth to the lth dependent variables, where $k \leq l$. These observations are at times $t_1 \leq t_2 \leq \cdots \leq t_T$. Our objective is to find values of the parameters α_1, $\alpha_2, \ldots, \alpha_m$ such that we minimize the sum of squared deviations,

$$S = \sum_{j=k}^{l} \sum_{r=1}^{T} [x_j(t_r) - b_{jr}]^2. \tag{4.3}$$

One approach to solving this system identification problem is quasilinearization, which has been used in the field of control engineering since the 1950s (References 1–3, 9).

Since the solution of the system of equations in (4.1) is a nonlinear function of the parameters $\alpha_1, \alpha_2, \ldots, \alpha_m$, we first linearize the system via a Taylor series expansion. With the linear system we know that the parameters $\alpha_1, \alpha_2, \ldots, \alpha_m$ will appear linearly in the general form of the solution. Because quasilinearization is a successive approximation scheme, we select $\alpha_1^1, \alpha_2^1, \ldots, \alpha_m^1$ as the initial approximations to the minimizing values of the parameters; the functions $x_1^1(t), x_2^1(t), \ldots, x_n^1(t)$ are determined from the equations

$$\dot{x}_i^1 = f^i(x_1^1, x_2^1, \ldots, x_n^1; a_1^1, \alpha_2^1, \ldots, \alpha_m^1), \qquad 0 \leq t \leq t_T, \tag{4.4}$$

$$x_i^1(0) = \gamma_i, \qquad i = 1, 2, \ldots, n. \tag{4.5}$$

For the first approximation, $K = 1$, we merely adjoin the differential equations for p_i and q_{ij} below, integrate, and solve for $\alpha_1^2, \alpha_2^2, \ldots, \alpha_m^2$. For approximations 2 through K, $K > 2$, the linearized system has the form

$$\begin{aligned} \dot{x}_i^{L+1} = {} & f^i(x_1^L, x_2^L, \ldots, x_n^L; \alpha_1^L, \alpha_2^L, \ldots, \alpha_m^L) \\ & + \sum_{j=1}^{n} (x_j^{L+1} - x_j^L) f_{x_j}^i(x_1^L, x_2^L, \ldots, x_n^L; \alpha_1^L, \alpha_2^L, \ldots, \alpha_m^L) \\ & + \sum_{j=1}^{m} (\alpha_j^{L+1} - \alpha_j^L) f_{\alpha_j}^i(x_1^L, x_2^L, \ldots, x_n^L; \alpha_1^L, \alpha_2^L, \ldots, \alpha_m^L), \end{aligned} \tag{4.6}$$

$$\begin{aligned} x_i^{L+1}(0) &= \gamma_i, \\ i &= 1, 2, \ldots, n, \\ L &= 1, 2, \ldots, K-1. \end{aligned} \tag{4.7}$$

Assuming that the minimizing values of the parameters in the Kth approximation are already determined, the next linearized system will be

$$\begin{aligned} \dot{x}_i^{K+1} = {} & f^i(x_1^K, x_2^K, \ldots, x_n^K; \alpha_1^K, \alpha_2^K, \ldots, \alpha_m^K) \\ & + \sum_{j=1}^{n} (x_j^{K+1} - x_j^K) f_{x_j}^i(x_1^K, x_2^K, \ldots, x_n^K; \alpha_1^K, \alpha_2^K, \ldots, \alpha_m^K) \\ & + \sum_{j=1}^{m} (\alpha_j^{K+1} - \alpha_j^K) f_{\alpha_j}^i(x_1^K, x_2^K, \ldots, x_n^K; \alpha_1^K, \alpha_2^K, \ldots, \alpha_m^K), \end{aligned} \tag{4.8}$$

$$x_i^{K+1}(0) = \gamma_i, \qquad i = 1, 2, \ldots, n. \tag{4.9}$$

The general form of the solution to this system is

$$x_i^{K+1} = p_i(t) + \sum_{j=1}^{m} \alpha_j^{K+1} q_{ij}(t), \qquad i = 1, 2, \ldots, n, \tag{4.10}$$

where $p_i(t)$ is a particular solution, and $q_{ij}(t)$ forms another particular solution. The functions $p_i(t)$ and $q_{ij}(t)$ are the solutions of the initial value problems

$$\begin{aligned}\dot{p}_i = {} & \sum_{j=1}^{n} f_{x_j}^i(x_1^K, x_2^K, \ldots, x_n^K; \alpha_1^K\ \alpha_2^K, \ldots, \alpha_m^K)p_j \\ & + f^i(x_1^K, x_2^K, \ldots, x_n^K; \alpha_1^K, \alpha_2^K, \ldots, \alpha_m^K) \\ & - \sum_{j=1}^{n} x_j^K f_{x_j}^i(x_1^K, x_2^K, \ldots, x_n^K; \alpha_1^K, \alpha_2^K, \ldots, \alpha_m^K) \\ & - \sum_{j=1}^{m} \alpha_j^K f_{\alpha_j}^i(x_1^K, x_2^K, \ldots, x_n^K; \alpha_1^K, \alpha_2^K, \ldots, \alpha_m^K), \\ & \qquad i = 1, 2, \ldots, n, \end{aligned} \tag{4.11}$$

$$p_i(0) = \gamma_i, \tag{4.12}$$

and

$$\begin{aligned}\dot{q}_{ij} = {} & \sum_{r=1}^{n} f_{x_r}^i(x_1^K, x_2^K, \ldots, x_n^K; \alpha_1^K, \alpha_2^K, \ldots, \alpha_m^K)q_{rj} \\ & + f_{\alpha_j}^i(x_1^K, x_2^K, \ldots, x_n^K; \alpha_1^K, \alpha_2^K, \ldots, \alpha_m^K), \end{aligned} \tag{4.13}$$

$$\begin{aligned} q_{ij}(0) &= 0, \\ i &= 1, 2, \ldots, n, \\ j &= 1, 2, \ldots, m. \end{aligned} \tag{4.14}$$

Equation (4.11) involves all the known initial conditions and forcing terms in the equation (4.8), and equation (4.13) involves the derivatives of the right-hand sides of the original differential equations with respect to the parameters.

At this point assume that all the approximations up to the parameters $\alpha_1^K, \alpha_2^K, \ldots, \alpha_m^K$ have been determined. We use a numerical technique called the "simultaneous calculation of approximations" method. This technique allows us to increase the accuracy of the numerical results and save memory space. Instead of calculating and storing the functions $x_1^K(t), x_2^K(t), \ldots, x_n^K(t)$ at every approximation, equations (4.4), (4.6), (4.11), and (4.13), subject to the initial conditions in equations (4.5), (4.7), (4.12), and (4.14) are integrated simultaneously.

Now that the functions $p_i(t)$ and $q_{ij}(t)$ have been computed, everything in the general solution in equation (4.10) is known except the new approximations to the unknown parameters $\alpha_1^{K+1}, \alpha_2^{K+1}, \ldots, \alpha_m^{K+1}$. Recall that the function to be minimized is given in equation (4.3). The values of the parameters $\alpha_1^{K+1}, \alpha_2^{K+1}, \ldots, \alpha_m^{K+1}$ are to be chosen such that the sum of squared deviations is minimized.

The function that we actually minimize is

$$Q = \sum_{j=k}^{l} \sum_{r=1}^{T} \left[p_j(t_r) + \sum_{s=1}^{m} \alpha_s^{K+1} q_{js}(t_r) - b_{jr} \right]^2, \tag{4.15}$$

where Q is an approximation to S in equation (4.3).

To find the values of the parameters $\alpha_1^{K+1}, \alpha_2^{K+1}, \ldots, \alpha_m^{K+1}$ which minimize Q, the partial derivatives of the function Q with respect to each of the unknown parameters are set equal to zero,

$$\partial Q / \partial \alpha_j^{K+1} = 0, \qquad j = 1, 2, 3, \ldots, m, \tag{4.16}$$

which yields a system of linear algebraic equations, the normal equations,

$$\sum_{j=k}^{l} \sum_{r=1}^{T} \left[p_j(t_r) + \sum_{s=1}^{m} \alpha_s^{K+1} q_{js}(t_r) - b_{jr} \right] q_{jh}(t_r) = 0,$$
$$h = 1, 2, 3, \ldots, m. \tag{4.17}$$

This linear algebraic system of equations is to be solved for the new approximation to the parameters.

Simplifying the linear algebraic system in equation (4.17), we obtain

$$\sum_{h=1}^{m} \alpha_h^{K+1} \sum_{r=1}^{T} \sum_{j=k}^{l} q_{jh}(t_r) q_{jd}(t_r) = \sum_{r=1}^{T} \sum_{j=k}^{l} [b_{jr} - p_j(t_r)] q_{jd}(t_r),$$
$$d = 1, 2, 3, \ldots, m. \tag{4.18}$$

Assuming the linear algebraic system is not singular or ill conditioned, it can be solved for the new approximations to the unknown parameters, $\alpha_1^{K+1}, \alpha_2^{K+1}, \ldots, \alpha_m^{K+1}$.

The process is then repeated, if necessary, with K being replaced by $K + 1$. If the initial approximations of the parameters $\alpha_1^1, \alpha_2^1, \ldots, \alpha_m^1$ are sufficiently close to the minimizing values of the parameters, only a few iterations will be needed for convergence.

Quasilinearization is a quadratically convergent technique. If convergence occurs, with each new approximation the number of correct digits is doubled, approximately.

4.2. Fast and Efficient Evaluation of Derivatives (FEED)

To evaluate the solutions of the system of ordinary differential equations, we need to have the numerical values of the right-hand sides of these equations. Forming these analytical partial derivatives can sometimes be tedious and difficult. However, what we really need is a knowledge of the numerical values of the derivatives, not their explicit analytical expressions. Consequently in this chapter we apply a recent development, FEED (References 5–8), to evaluate the first derivatives of these equations. This method was first presented in Reference 8 and was further developed and used in various applications (References 5–7). This method can be used in any problem where values of partial derivatives are required; the user specifies the form of the function, and the derivatives are evaluated automatically.

It is easiest to review the technique by means of an example. Let

$$z = x + y e^{x}. \tag{4.19}$$

We wish to evaluate the numerical values of z, z_x, z_y, and z_{xx}, for given values of x and y. First we write the function in a sequential fashion, as in the first column of Table 4.1. Note that each one of the steps 1–5 is a very simple operation on one or two variables only. The second and third columns of Table 4.1 are obtained by differentiating both sides of the entries in the first column of the table with respect to x and y. The last column is obtained by taking the second derivative of the first column with respect to x.

Clearly the left-hand sides in row 5 are the desired quantities. Now let us consider each row of the table. To carry out the arithmetic by computer we need three types of subroutines. For rows 1 and 2 we need an "initializing" or "vectorizing" routine which takes as inputs the values of x and y and outputs two vectors with the components A, A_x, A_y, A_{xx}, and B, B_x, B_y, B_{xx}. This subroutine is denoted by LIN in our FORTRAN program.

Row 3 requires a subroutine that, given the values A, A_x, A_y, and A_{xx}, computes the value of the special function $C = e^{A}$, the value of its derivatives with respect to x and y, C_x, and C_y, and the value of its second-order derivative with respect to x, C_{xx}. These values are the components of a four-dimensional vector in the order, C, C_x, C_y, and C_{xx} (in the FORTRAN program at the end of the chapter this would be handled by subroutine CONTTU).

Row 4 requires a subroutine which, given B, and C, computes the value of their product $D = BC$, and the values of the product's derivatives with respect to x, and y, and the value of its second-order derivative with respect to x. These values are the components of a four-dimensional vector with components D, D_x, D_y, and D_{xx}. This routine is denoted MULT in the FORTRAN program.

Table 4.1. FEED

	Function	$\partial/\partial x$	$\partial/\partial y$	$\partial^2/\partial x^2$
1.	$A = x$	$A_x = 1$	$A_y = 0$	$A_{xx} = 0$
2.	$B = y$	$B_x = 0$	$B_y = 1$	$B_{xx} = 0$
3.	$C = e^A$	$C_x = A_x e^A$	$C_y = A_y e^A$	$C_{xx} = A_{xx} e^A + (A_x)^2 e^A$
4.	$D = BC$	$D_x = B_x C + BC_x$	$D_y = B_y C + BC_y$	$D_{xx} = B_{xx} C + 2B_x C_x + BC_{xx}$
5.	$Z = A + D$	$Z_x = A_x + D_x$	$Z_y = A_y + D_y$	$Z_{xx} = A_{xx} + D_{xx}$

Finally the last row of Table 4.1 is the addition of two variables A and D that have already been computed in the preceding rows. Thus we need a subroutine which, given the values of A, A_x, A_y, and A_{xx} and D, D_x, D_y, and D_{xx}, evaluates the sum $Z = A + D$, its first derivatives with respect to x and y, and its second-order derivative with respect to x. These values, Z, Z_x, Z_y, and Z_{xx}, are the components of a four-dimensional vector. Subroutine ADD in the FORTRAN listing performs this operation.

This ends the entire sequential evaluation of the function z, its derivatives with respect to x and y, and its second derivative with respect to x. Then, for these general calculus subroutines to perform their tasks, we need a special kind of routine which basically functions as an organizer of the FEED library. This subroutine calls the vectorizing, and the calculus routines in order to do the FEED procedure. This routine is denoted FUN in our FORTRAN listing, and in Section 4.3 its function is discussed in more detail.

This example plus study of the examplė of Section 4.4 and subroutine FUN make the formalism clear.

4.3. Computer Program

The FORTRAN program is written for a system of n ordinary differential equations with m parameters in the model to be estimated. This program consists of a main program and subroutines of several types. One group deals with the integration of the differential equations. A second type is for solving the linear algebraic equations. The third group carries out the FEED procedures for calculating the numerical values of the derivatives.

The main program sets the initial approximations for the parameters, and performs the initializing of the differential equations. The number of observations, the step size in time, the first and the last variables to be observed, and the number of iterations are specified here. The program is dimensioned for a maximum of five iterations.

The maximum number of ordinary differential equations to be simultaneously integrated is $(K + 1 + m)n$, where n is the number of dependent variables, m is the number of unknown parameters, and K is the number of iterations to be obtained. The T vector must be dimensioned at least ten times the order of the differential system in the last iteration.

The matrix XX with 10 rows and 5 columns is the matrix of the dependent variables. The first column of this matrix consists of the initial approximations, the functions $x_1^1(t), \ldots, x_{10}^1(t)$. These functions are the solution of the original system of ordinary differential equations (4.1), given the initial approximations to the parameters. Each additional column of this matrix represents the next approximation to the solution functions $x_1(t), \ldots, x_{10}(t)$. The matrix ALFA has five rows and six columns. The first column of matrix ALFA consists of the initial approximations to the parameters, $\alpha_1^1, \ldots, \alpha_5^1$. Columns 2–6 hold the improved approximations in iterations 1–5.

To solve for the minimizing values of the parameters $\alpha_1, \ldots, \alpha_5$ the matrix of coefficients CMAT and the right-hand side vector RHS are formed. In every iteration, when we pass any point of observation, the elements of the matrix CMAT and the vector RHS are updated. They then are placed into the matrix A. These elements are some combination of the functions $p_i(t)$ and $q_{ij}(t)$, and are the coefficients of the parameters in the sum of squared deviations given in equation (4.18). Then the system of linear algebraic equations for the minimizing values of the parameters is solved using Gaussian elimination performed on matrix A. Subroutine GAUSS is used to carry out this operation through a call from the main program.

Subroutine GEDAT generates the data. The information about the number of variables, and the first and the last variables to be observed, is communicated from the main program to GEDAT through a common statement. The matrix of observations B is the final product of this subroutine. This matrix is dimensioned at 10 by 100. The number of rows is the maximum number of dependent variables which can be observed. We can have up to 100 times of observation for each variable. The element b_{ij} of matrix B is the observation at time j of the ith dependent variable.

One category of subroutine in the program deals with integration of the differential equations. This group consists of subroutines DAUX, DINT1, and DINT2.

Subroutine DAUX basically forms the right-hand sides of the differential equations. The inputs to this subroutine are the T vector, the number of differential equations, time, and the integration step size. The value of KKK, the number of approximations completed, is communicated to DAUX through a common statement. The vector PD and matrices QD and XXD contain values of the right-hand sides of the differential equations (4.11), (4.13), and (4.4) and (4.8).

The initial conditions are placed in the first NDE (number of differential equations) positions of the T vector in the main program. We take these values and map them into P, Q, and X because we want to use them for evaluating PD, QD, and XXD. The values of the derivatives are computed in subroutine DAUX. Then, these derivatives are placed in positions NDE+1 up to 2(NDE) in the T vector. We always fill in the T region in the order P, Q, and XX. To compute the derivatives subroutine DAUX makes several calls to subroutine FUN, which is one of the FEED subroutines discussed next.

Subroutine DINT2 carries out a standard Adams–Moulton start for the numerical integration followed by Runge–Kutta continuation, both of fourth order.

The third group of subroutines carries out the FEED procedure for calculating the numerical values of the needed derivatives. To implement FEED, routines of four types are needed. First we need to supply a routine which calculates the "vectorized form" of the variables $x_1, \ldots, x_n$, and $\alpha_1, \ldots, \alpha_m$. Subroutine LIN is used to vectorize these variables. The inputs to this subroutine are the vectors XXX, which can contain up to ten dependent variables in the system of ordinary differential equations, and ALFA1, which can contain up to five unknown parameters. This routine produces $m + n$ vectors of dimension $1 + n + m$, where n is the number of dependent variables, and m is the number of parameters. Each of these vectors consists of the value of the variable, a one in the appropriate place for the derivative of the variable with respect to itself, and zeros elsewhere for the derivatives of the variable with respect to the other variables. As an example, the vectorized form of the variable x_2 is the vector $X2$, which has the form

$$(x_2, 0, 1, 0, 0, 0, 0, 0, 0, 0, 0, 0, 0, 0, 0, 0).$$

The second type of subroutine in the FEED procedure is the type used for functions of two variables. These are only five functions of two variables used: addition, subtraction, multiplication, division, and raising a variable to a variable power. In the illustrative program we only need addition, and multiplication of two variables. Subroutines ADD and MULT are written for these two operations.

Each function of one variable requires a special routine. Examples are $\log x$, $\exp(x)$, and so on. In our program subroutines CONST and CONTTU are functions of one variable. Subroutine CONST creates a 16-dimensional vector with the first element equal to a constant number and the rest of the elements equal to zero (the derivative of the constant with respect to x_1, $\ldots$, x_{10}, and $\alpha_1, \ldots, \alpha_5$). Subroutine CONTTU is used to create a vectorized form of a constant raised to a variable power.

Finally we need a routine to organize the operations of the other FEED subroutines to create the functions on the right-hand sides of the differential equations. Subroutine FUN is basically written for this purpose. It was mentioned earlier that DAUX calls FUN to carry out the FEED procedure to calculate the numerical values of the right-hand sides of the differential equations. Before each call to subroutine FUN, for each iteration K, the Kth column of matrix XX is transferred into the vector XXX, and the Kth column of the matrix ALFA is transferred into the vector ALFA1. These vectors are inputs to the subroutine FUN. Time is also input to this subroutine. The output of this routine is the n by $1 + n + m$ dimensional matrix F.

To implement the FEED procedure FUN calls LIN to vectorize each element of the vectors XXX and ALFA1. Then other subroutines in the FEED library are called. These routines are used for operations on functions of two variables or functions of one variable discussed earlier. The result, which is in a vector form, is placed in the appropriate row of the matrix F.

One objective in writing this general program has been to minimize the number of modifications the user has to make. In the main program the user specifies the number of dependent variables, the number of parameters to be estimated, the number of iterations, the number of the first and the last variables to be observed, the number of observations on each variable, the integration step size, the initial guesses of the parameters of the model that are being estimated, and the initial conditions for the differential equations.

Among the FEED subroutines, FUN is the only one that has to be modified according to the user's model. We have not included a wide variety of subroutines for the functions of one variable and two variables; in fact, we provide only the ones needed for our sample model. However, some of these routines are available in other parts of the book. In addition, it is possible to do some operations by using two or more of the available routines. For example, subtraction of two variables can be done by using subroutines MULCON and ADD.

To solve the system of linear algebraic equations, subroutine GAUSS is used. This routine can be replaced by a more general type of subroutine which is more powerful in dealing with ill-conditioned systems. In cases where the columns of matrix CMAT are not linearly independent, the ordinary inverse of matrix CMAT does not exist. The Moore–Penrose or generalized inverse of the matrix CMAT may be used. We have written a FORTRAN program which will calculate the generalized inverse; it will be included in future versions as an option.

4.4. Numerical Results

In this section we consider estimating the parameters of a system of ordinary differential equations, given a finite number of observations. The model which is considered, the Weibull distribution, is semi-Markov in nature and is used to predict target acquisition probability, in an image, as a function of search time.

The problem of fitting this Markov process to data is determining parameters α_1 and α_2. These parameters are in the system of two ordinary differential equations

$$\dot{x}_1 = -\alpha_1\alpha_2 t^{\alpha_2-1}x_1, \tag{4.20}$$

$$\dot{x}_2 = \alpha_1\alpha_2 t^{\alpha_2-1}x_1, \tag{4.21}$$

and the initial conditions

$$x_1(0) = 1, \tag{4.22}$$

$$x_2(0) = 0. \tag{4.23}$$

The probability that search is terminated by time t is $x_2(t)$, and the probability that search is terminated at time t or later is $x_1(t)$, $t > 0$.

There are five observations on the variable x_2, $x_2(2.0) = 0.2$, $x_2(2.6) = 0.4$, $x_2(3.7) = 0.6$, $x_2(5.2) = 0.8$, and $x_2(9.2) = 1.0$. These observations are stored in the second row of the matrix B in columns one through five in subroutine GEDAT. Our objective is simultaneously to estimate the values of the parameters α_1 and α_2 that give the best fit of the model to data, using the technique discussed in Section 4.1.

We start with initial estimates of 0.1 and 0.2 for α_1 and α_2. These initial approximations are determined by some statistical considerations

Table 4.2. Numerical Results

α_1	α_2
0.1000000000 D + 00	0.2000000000 D + 01
0.8258009200 D − 01	0.1784188439 D + 01
0.7159096166 D − 01	0.1921153682 D + 01
0.7340895078 D − 01	0.1907411980 D + 01
0.7340937929 D − 01	0.1907548421 D + 01
0.7341071451 D − 01	0.1907534209 D + 01

concerning the analytical solution of the equation (4.19) (Reference 4). Table 4.2 reports the rate of convergence. After five iterations both α_1 and α_2 have converged to six significant digits.

4.5. Program Listing

The FORTRAN program is currently dimensioned to handle the estimation problem for a system of up to ten differential equations and five parameters. For a bigger system, the program has to be redimensioned. Also, subroutine GAUSS, which performs Gaussian elimination for the system of linear algebraic equations, can be replaced by another routine which obtains Moore–Penrose generalized inverses.

```
C     MAIN PROGRAM FOR QL. WITH SCAM AND FEED FOR A SYSTEM OF D.EQS.
      IMPLICIT REAL*8(A-H,O-Z)
      COMMON ALFA
      COMMON KKK, NVAR,NALFA,NOBS,NISBO
      COMMON/A1/IFLAG
      DIMENSION T(1100),XX(10,5),ALFA(5,6),B(10,100),A(10,11)
      DIMENSION RHS(5),P(10),Q(10,5),CMAT(5,5),GAMA(10),INDEX(5)
      TMIN-0.0D+00
      IFLAG=1
      NVAR=2
      NALFA=2
      NITER=5
      NOBSF=2
      NOBSL=2
      NOBS=5
      H=0.01D+00
      ALFA(1,1)=0.10D+00
      ALFA(2,1)=2.0D+00
      WRITE(6,100)(ALFA(II,1),II=1,NALFA)
      GAMA(1)=1.0D+00
      GAMA(2)=0.0D+00
      CALL GEDAT(E,H)
C     DO THE ITERATIONS
      DO 1000 KK=1,NITER
      KKK=KK
      NDE=(KKK+1+NALFA)*NVAR
C     INITIALIZING CMAT AND RHS
      DO 21 K=1,NALFA
      DO 23 JJ=1,NALFA
      CMAT(K,JJ)=0.0D+00
23    CONTINUE
      RHS(K)=0.0D+00
21    CONTINUE
C     INITIALIZING P
      DO 25 I=1,NVAR
      P(I)=GAMA(I)
25    CONTINUE
C     INITIALIZING Q
```

```
      DO 27 I=1,NVAR
      DO 29 J=1,NALFA
      Q(I,J)=0.0D+00
 29   CONTINUE
 27   CONTINUE
 C    INITIALIZING XX
      DO 31 J=1,KKK
      DO 33 I=1,NVAR
      XX(I,J)=GAMA(I)
 33   CONTINUE
 31   CONTINUE
 C    MAP P INTO T
      L=0
      DO 35 I=1,NVAR
      L=L+1
      T(L)=P(I)
 35   CONTINUE
 C    MAP Q INTO T
      DO 37 J=1,NALFA
      DO 39 I=1,NVAR
      L=L+1
      T(L)=Q(I,J)
 39   CONTINUE
 37   CONTINUE
 C    MAP XX INTO T
      DO 41 J=1,KKK
      DO 43 I=1,NVAR
      L=L+1
      T(L)=XX(I,J)
 43   CONTINUE
 41   CONTINUE
 C    INITIALIZE X
      X=TMIN
      CALL DINT1(T,NDE,X,H)
 C    CARRY OUT THE INTEGRATION
      INDEX(1)=200
      INDEX(2)=60
      INDEX(3)=110
      INDEX(4)=150
      INDEX(5)=390
      DO 45 JJ=1,NOBS
      NISBO=INDEX(JJ)
      DO 71 JI=1,NISBO
      CALL DINT2(T,NDE,X,H)
 71   CONTINUE
 C    MAP CURRENT T INTO P, AND G
      L=0
      DO 47 I=1,NVAR
      L=L+1
      P(I)=T(L)
 47   CONTINUE
      DO 49 J=1,NALFA
      DO 51 I=1,NVAR
      L=L+1
      Q(I,J)=T(L)
 51   CONTINUE
```

```
49      CONTINUE
C       BUMP UP CMAT
        DO 53 J1=1,NALFA
        DO 55 K=1,NALFA
        SUM=0.0D+00
        DO 57 I=NOBSF,NOBSL
        PROD=Q(I,J1)*Q(I,K)
        SUM=SUM+PROD
57      CONTINUE
        CMAT(K,J1)+CMAT(K,J1)+SUM
55      CONTINUE
53      CONTINUE
C       BUMP UP RHS
        DO 59 K=1,NALFA
        SUM=0.0D+00
        DO 61 I=NOBSF,NOBSL
        PROD=(B(I,JJ)-P(I))*Q(I,K)
        SUM=SUM+PROD
61      CONTINUE
        RHS(K)=RHS(K)+SUM
59      CONTINUE
45      CONTINUE
C       FORM MATRIX A
        DO 63 II=1,NALFA
        DO 65 IJ=1,NALFA
        A(II,IJ)=CMAT(II,IJ)
65      CONTINUE
63      CONTINUE
        NP=NALFA+1
        DO 67 II=1,NALFA
        A(II,NP)=RHS(II)
67      CONTINUE
        CALL GAUSS(A,NALFA)
        KKP1=KKK+1
        DO 69 II=1,NALFA
        ALFA(II,KKP1)=A(II,NP)
69      CONTINUE
        WRITE(6,100)(ALFA(II,KKP1),II=1,NALFA)
1000    CONTINUE
100     FORMAT(1X,5D24.10)
        STOP
        END

        SUBROUTINE GEDAT(B,H)
        IMPLICIT REAL*B(A-H,O-Z)
        COMMON ALFA(5,6)
        COMMON KKK,NVAR,NALFA,NOBS,NISBO
        DIMENSION B(10,100)
        B(2,1)=0.2D+00
        B(2,2)=0.4D+00
        E(2,3)=0.6D+00
        B(2,4)=0.8D+00
        B(2,5)=1.0D+00
        RETURN
        END
```

```
      SUBROUTINE FUN(XXX,ALFA1,TIME,F)
      IMPLICIT REAL*8(A-H,O-Z)
      COMMON ALFA(5,6)
      COMMON KKK,NVAR,NALFA,NOBS,NISBO
      DIMENSION XXX(10),ALFA1(5),F(10,16)
      DIMENSION Q(16),T(16),W(16),E(16),F1(16),G(16),H(16),P(16)
      DIMENSION X1(16),X2(16),X3(16),X4(16),X5(16),X6(16)
      DIMENSION X7(16),X8(16),X9(16),X10(16),A1(16),A2(16),A3(16)
      DIMENSION A4(16),A5(16)
      CALL LIN(XXX,ALFA1,X1,X2,X3,X4,X5,X6,X7,X8,X9,X10,A1,A2,A3,
     $A4,A5)
      CON=-1.0D+00
      CALL MULCON(CON,A1,E)
      CALL MULT(E,A2,F1)
      CALL MULT(F1,X1,G)
      CALL CONST(CON,H)
      CALL ADD(A2,H,P)
      CON=TIME
      CALL CONTTU(CON,P,Q)
      CALL MULT(G,Q,T)
      MM=1+NVAR+NALFA
      DO 105 J=1,MM
      F(1,J)=T(J)
105   CONTINUE
      CON=-1.0D+00
      CALL MULCON(CON,T,W)
      DO 107 J=1,MM
      F(2,J)=W(J)
107   CONTINUE
      RETURN
      END

      SUBROUTINE LIN(XXX,ALFA1,X1,X2,X3,X4,X5,X6,X7,X8,X9,
     $X10,A1,A2,A3,A4,A5)
      IMPLICIT REAL*8(A-H,O-Z)
      COMMON ALFA(5,6)
      COMMON KKK,NVAR,NALFA,NOBS,NISBO
      COMMON/A1/IFLAG
      DIMENSION XXX(10),ALFA1(5)
      DIMENSION X1(16),X2(16),X3(16),X4(16),X5(16),X6(16)
      DIMENSION X7(16),X8(16),X9(16),X10(16)
      DIMENSION A1(16),A2(16),A3(16),A4(16),A5(16)
      MM=1+NVAR+NALFA
      ZERO=0.0D+00
      ONE=1.0D+00
      IF(IFLAG.EQ.2) GO TO 150
      GO TO (31,32,33,34,35,36,37,38,39,40) ,NVAR
40    DO 50 I=1,MM
50    X10(I)=ZERO
      X10(11)=ONE
39    DO 51 I=1,MM
51    X9(I)=ZERO
      X9(10)=ONE
38    DO 52 I=1,MM
52    X8(I)=ZERO
      X8(9)=ONE
```

```
37      DO 53 I=1,MM
53      X7(I)=ZERO
        X7(8)=ONE
36      DO 54 I=1,MM
54      X6(I)=ZERO
        X6(7)=ONE
35      DO 55 I=1,MM
55      X5(I)=ZERO
        X5(6)=ONE
34      DO 56 I=1,MM
56      X4(I)=ZERO
        X4(5)=ONE
33      DO 57 I=1,MM
57      X3(I)=ZERO
        X3(4)=ONE
32      DO 58 I=1,MM
58      X2(I)=ZERO
        X2(3)=ONE
31      DO 59 I=1,MM
59      X1(I)=ZERO
        X1(2)=ONE
        GO TO(41,42,43,44,45) ,NALFA
45      DO 60 I=1,MM
60      A5(I)=ZERO
        A5(NVAR+6)=ONE
44      DO 61 I=1,MM
61      A4(I)=ZERO
        A4(NVAR+5)=ONE
43      DO 62 I=1,MM
62      A3(I)=ZERO
        A3(NVAR+4)=ONE
42      DO 63 I=1,MM
63      A2(I)=ZERO
        A2(NVAR+3)=ONE
41      DO 64 I=1,MM
64      A1(I)=ZERO
        A1(NVAR+2)=ONE
        IFLAG=2
150     CONTINUE
        GO TO(71,72,73,74,75,76,77,78,79,80),NVAR
80      X10(1)=XXX(10)
79      X9(1)=XXX(9)
78      X8(1)=XXX(8)
77      X7(1)=XXX(7)
76      X6(1)=XXX(6)
75      X5(1)=XXX(5)
74      X4(1)=XXX(4)
73      X3(1)=XXX(3)
72      X2(1)=XXX(2)
71      X1(1)=XXX(1)
        GO TO(81,82,83,84,85  ),NALFA
85      A5(1)=ALFA1(5)
84      A4(1)=ALFA1(4)
83      A3(1)=ALFA1(3)
82      A2(1)=ALFA1(2)
81      A1(1)=ALFA1(1)
```

```
      RETURN
      END

      SUBROUTINE ADD(A,B,C)
      IMPLICIT REAL*8(A-H,O-Z)
      COMMON ALFA(5,6)
      COMMON KKK,NVAR,NALFA,NOBS,NISBO
      DIMENSION A(16),B(16),C(16)
      MM=1+NVAR+NALFA
      DO 35 I=1,MM
      C(I)=A(I)+B(I)
35    CONTINUE
      RETURN
      END

      SUBROUTINE MULT(F,A,G)
      IMPLICIT REAL*8(A-H,O-Z)
      COMMON ALFA(5,6)
      COMMON KKK,NVAR,NALFA,NOBS,NISBO
      DIMENSION F(16),A(16),G(16)
      G(1)=F(1)*A(1)
      DO 25 I=1,NVAR
      INDEX=I+1
      G(INDEX)=F(INDEX)*A(1)+F(1)*A(INDEX)
25    CONTINUE
      NX=1+NVAR
      DO 27 J=1,NALFA
      LM=J+NX
      G(LM)=F(LM)*A(1)+F(1)*A(LM)
27    CONTINUE
      RETURN
      END

      SUBROUTINE CONST(CON,E)
      IMPLICIT REAL*8(A-H,O-Z)
      COMMON ALFA(5,6)
      COMMON KKK,NVAR,NALFA,NOBS,NISBO
      DIMENSION E(16)
      MM=1+NVAR+NALFA
      E(1)=CON
      DO 35 I=2,MM
      E(I)=0.CD+00
35    CONTINUE
      RETURN
      END

      SUBROUTINE CONTTU(C,U,Z)
      IMPLICIT REAL*8(A-H,O-Z)
      COMMON ALFA(5,6)
      COMMON KKK,NVAR,NALFA,NOBS,NISBO
      DIMENSION U(16), Z(16)
      MM=1+NVAR+NALFA
      IF(C.EQ.0.0D+00) GO TO 47
      AA=DLOG(C)
      Z(1)=DEXP(U(1)*AA)
      BB=AA*Z(1)
```

```
      DO 17 I=2,MM
      Z(I)=BB*U(I)
17    CONTINUE
      GO TO 67
47    DO 57 I=1,MM
57    Z(I)=0.0D+00
67    RETURN
      END

      SUBROUTINE MULCON(CON,C,S)
      IMPLICIT REAL*8(A-H,O-Z)
      COMMON ALFA(5,6)
      COMMON KKK,NVAR,NALFA,NOBS,NISBO
      DIMENSION C(16),S(16)
      MM=1+NVAR+NALFA
      DO 35 I=1,MM
      S(I)=CON*C(I)
35    CONTINUE
      RETURN
      END

      SUBROUTINE GAUSS(A,NALFA)
      IMPLICIT REAL*8(A-H,O-Z)
      DIMENSION A(10,11)
      NP1=NALFA+1
      EPS=1.0D-05
      DO 30 K=1,NALFA
      PIVOT=A(K,K)
      IF(DABS(PIVOT).LT.EPS)WRITE(6,103)
      DO 17 J=K,NP1
      A(K,J)=A(K,J)/PIVOT
17    CONTINUE
103   FORMAT(1X.'SMALL PIVOT')
      DO 25 I=1,NALFA
      IF(I.EQ.K)GO TO 25
      AIK=A(I,K)
      DO 20 J=K,NP1
      A(I,J)=A(I,J)-AIK*A(K,J)
20    CONTINUE
25    CONTINUE
30    CONTINUE
      RETURN
      END

      SUBROUTINE DAUX(T,NDE,X,H)
      IMPLICIT REAL*8(A-H,O-Z)
      COMMON ALFA
      COMMON KKK,NVAR,NALFA,NOBS,NISBO
      DIMENSION T(1100),P(10),Q(10,5),XX(10,5),PD(10),QD(10,5)
      DIMENSION XXD(10,5),F(10,16),XXX(10),ALFA1(5),ALFA(5,6)
C     DICTIONARY
      L=0
      DO 5 I=1,NVAR
      L=L+1
      P(I)=T(L)
5     CONTINUE
```

```
      DO 6 JJ=1,NALFA
      DO 7 I=1,NVAR
      L=L+1
      Q(I,JJ)=T(L)
7     CONTINUE
6     CONTINUE
      DO 8 JJJ=1,KKK
      DO 9 I=1,NVAR
      L=L+1
      XX(I,JJJ)=T(L)
9     CONTINUE
8     CONTINUE
      DO 17 I=1,NVAR
      XXX(I)=XX(I,KKK)
17    CONTINUE
      DO 19 JJ=1,NALFA
      ALFA1(JJ)=ALFA(JJ,KKK)
19    CONTINUE
      TIME=X
      CALL FUN(XXX,ALFA1,TIME,F)
C     WRITE(6,400) XXX(1),XXX(2)
C     WRITE(6,400) ALFA1(1),ALFA1(2)
C     WRITE(6,400) TIME
C     WRITE(6,400)(F(1,I),I=1,26)
C     WRITE(6,400)(F(2,I),I=1,26)
      DO 33 I=1,NVAR
      PD(I)=F(I,1)
      DO 21 LL=1,NVAR
      LLP=LL+1
      PROD=F(I,LLP)*P(LL)
      PD(I)=PD(I)+PROD
21    CONTINUE
      DO 23 LL=1,NVAR
      LLP=LL+1
      PROD=XX(LL,KKK)*F(I,LLP)
      PD(I)=PD(I)-PROD
23    CONTINUE
      NP=NVAR+1
      DO 25 NR=1,NALFA
      NPP=NP+NR
      PROD=ALFA(NR,KKK)*F(I,NPP)
      PD(I)=PD(I)-PROD
25    CONTINUE
23    CONTINUE
      DO 35 I=1,NVAR
      DO 37 J=1,NALFA
      NP1=NVAR+1+J
      QD(I,J)=F(I,NP1)
37    CONTINUE
35    CONTINUE
      DO 39 I=1,NVAR
      DO 41 J=1,NALFA
      DO 43 L=1,NVAR
      LP1=L+1
      PROD=F(I,LP1)*Q(L,J)
      QD(I,J)=GD(I,J)+PROD
```

```
43      CONTINUE
41      CONTINUE
39      CONTINUE
        DO 45 I=1,NVAR
        XXX(I)=XX(I,1)
45      CONTINUE
        DO 47 J=1,NALFA
        ALFA1(J)=ALFA(J,1)
47      CONTINUE
        TIME=X
        CALL FUN(XXX,ALFA1,TIME,F)
        DO 49 I=1,NVAR
        XXD(I,1)=F(I,1)
49      CONTINUE
        IF(KKK.EQ.1) GO TO 67
        DO 51 MMM=2,KKK
        MMM1=MMM-1
        DO 53 I=1,NVAR
        XXX(I)=XX(I,MMM1)
53      CONTINUE
        DO 55 J=1,NALFA
        ALFA1(J)=ALFA(J,MMM1)
55      CONTINUE
        TIME=X
        CALL FUN(XXX,ALFA1,TIME,F)
        DO 57 I=1,NVAR
        XXD(I,MMM)=F(I,1)
57      CONTINUE
        DO 61 I=1,NVAR
        DO 59 L=1,NVAR
        LP1=L+1
        PROD=(XX(L,MMM)-XX(L,MMM1)*F(I,LP1)
        XXD(I,MMM)=XXD(I,MMM)+PROD
59      CONTINUE
61      CONTINUE
        DO 63 I=1,NVAR
        DO 65 NR=1,NALFA
        NRP=NR+1+NVAR
        PROD=(ALFA(NR,MMM)-ALFA(NR,MMM1))*F(I,NRP)
        XXD(I,MMM)=XXD(I,MMM)+PROD
65      CONTINUE
63      CONTINUE
51      CONTINUE
67      CONTINUE
C       DERIVATIVES NOW HAVE BEEN COMPUTED
C       MAP DERIVATIVES INTO T REGION
        L=NDE
        DO 71 I=1,NVAR
        L=L+1
        T(L)=PD(I)
71      CONTINUE
        DO 73 J=1,NALFA
        DO 75 I=1,NVAR
        L=L+1
        T(L)=QD(I,J)
75      CONTINUE
```

```
 73     CONTINUE
        DO 77 J=1,KKK
        DO 79 I=1,NVAR
        L=L+1
        T(L)=XXD(I,J)
 79     CONTINUE
 77     CONTINUE
 400    FORMAT(1X,5D24.10)
 300    FORMAT(1X,D24.10)
        RETURN
        END

        SUBROUTINE DINT1(T,N,X,H)
        IMPLICIT REAL*8(A-H,O-Z)
        COMMON /NSEQ/ N2,N3,N4,N5,N6,N7,N8,N9,NN,KFLAG,IND
        COMMON /HSEQ/ H2,H4,H24,R
        DIMENSION T(1)
C
C       CALC. CONSTANTS TO BE USED IN PROGRAM
        NN=N
        N2=N*2
        N3=N*3
        N4=N*4
        N5=N*5
        N6=N*6
        N7=N*7
        N8=N*8
        N9=N*9
        H2=H*0.5D+00
        H4=H2*0.5D+00
        H24=H/24.0D+00
        R=1.0D+00/6.0D+00
C       CALC. Y PRIME FOR INITIAL CONDITIONS
        CALL DAUX(T,N,X,H)
        DO 1 I=1,N
        N9I=I+N9
C       TEMPORARY STORAGE FOR Y
        N8I=I+N8
        NNI=I+NN
        T(N9I)=T(I)
C       STORE Y PRIME AT N-3 FOR USE IN A.M. INTEGRATION
      1 T(N8I)=T(NNI)
        KFLAG=0
        IND=0
        RETURN
        END

        SUBROUTINE DINT2(T,N,X,H)
        IMPLICIT REAL*8(A-H,O-H)
        COMMON /NSEQ/ N2,N3,N4,N5,N6,N7,N8,N9,NN,KFLAG,IND
        COMMON /HSEQ/ H2,H4,H24,R
        DIMENSION T(1)
C
C       IND=FLAG FOR R.K. INTEGRATION WHEN EQUAL OR LESS THAN 3
C       KFLAG=FLAG FOR STORING PAST DERIVATIVES FOR USE IN A.M. INTEGRATION
C
```

```
C                RUNGE-KUTTA INTEGRATION (6 STEPS AT H/2)
C       NOTE- TWO STEPS OF R.K. ARE DONE WITH EACH CALL TO THIS SUBROUTINE
C             SO THAT PRINTOUT POINTS WILL BE AT STEPS OF H.
C
      IND=IND+1
      KFLAG=KFLAG+1
      IF (IND-3) 20,20,12
   20 DO 9 K=1,2
C             CALC. K1
      DO 1 I=1,N
      N2I=I+N2
      NNI=I+NN
    1 T(N2I)=T(NNI)*H2
C             CALC. K2
C       STEP UP X
      X=X+H4
      DO 2 I=1,N
      N9I=I+N9
      N2I=I+N2
    2 T(I)=T(N9I)+0.5D+00 *T(N2I)
      CALL DAUX(T,N,X,H)
C       STORE K2
      DO 3 I=1,N
      N3I=I+N3
      NNI=I+NN
    3 T(N3I)=T(NNI)*H2
C             CALC. K3
      DO 4 I=1,N
      N9I=I+N9
      N3I=I+N3
    4 T(I)=T(N9I)+0.5D+00*T(N3I)
      CALL DAUX(T,N,X,H)
C       STORE K3
      DO 5 I=1,N
      N4I=I+N4
      NNI=I+NN
    5 T(N4I)=T(NNI)*H2
C             CALC. K4
C       STEP UP X
      X=X+H4
      DO 6 I=1,N
      N9I=I+N9
      N4I=I+N4
    6 T(I)=T(N9I)+T(N4I)
      CALL DAUX(T,N,X,H)
C       STORE K4
      DO 7 I=1,N
      N5I=I+N5
      NNI=I+NN
    7 T(N5I)=T(NNI)*H2
C       CALC. PREDICTED VALUE OF Y
      DO 8 I=1,N
      N9I=I+N9
      N2I=I+N2
      N3I=I+N3
      N4I=I+N4
```

```
      N5I=I+N5
      Y=T(N9I)+R*(T(N2I)+2.0D+00*T(N3I)+2.0D+00*T(N4I)+T(N5I))
C     STORE AS CURRENT VALUE OF Y
      T(I)=Y
C     STORE Y IN TEMPORARY STORAGE
    8 T(N9I)=T(I)
C     CALC. Y PRIME
      CALL DAUX(T,N,X,H)
C
C     STORE DERIVATIVES AS NEED FOR A.M.
      GO TO (10,11,9),KFLAG
C     STORE Y PRIME AT N-2
   10 DO 18 I=1,N
      N7I=I+N7
      NNI=I+NN
   18 T(N7I)=T(NNI)
      GO TO 9
C     STORE Y PRIME AT N-1
   11 DO 19 I=1,N
      N6I=I+N6
      NNI=I+NN
   19 T(N6I)=T(NNI)
    9 CONTINUE
      GO TO 17
C           ADAMS-MOULTON INTEGRATION
C
C     STORE CURRENT Y AND Y PRIME IN TEMPORARY STORAGE
   12 DO 13 I=1,N
      N9I=I+N9
      N2I=I+N2
      NNI=I+NN
      T(N9I)=T(I)
   13 T(N2I)=T(NNI)
C     CALC. PREDICTED VALUE OF Y
      DO 14 I=1,N
      N2I=I+N2
      N6I=I+N6
      N7I=I+N7
      N8I=I+N8
      N9I=I+N9
      YP=T(N9I)+H24*(55.0D+00*T(N2I)-59.0D+00*T(N6I)+37.0D+00*
     1  T(N7I)-9.0D+00*T(N8I))
C     STORE AS PREDICTED FUNCTIONAL VALUE OF Y AT X=X+H
   14 T(I)=YP
C     STEP UP X
      X=X+H
C     CALC. Y PRIME USING PREDICTED Y
      CALL DAUX(T,N,X,H)
C     CALC. CORRECTED Y
      DO 15 I=1,N
      NNI=I+NN
      N2I=I+N2
      N6I=I+N6
      N7I=I+N7
      N9I=I+N9
      YC=T(N9I)+H24*(9.0D+00*T(NNI)+19.0D+00*T(N2I)-5.0D+00*
     1  T(N6I)+T(N7I))
```

```
C       STORE AS NEW CURRENT VALUE OF Y
     15 T(I)=YC
C       CALC. Y PRIME TO BE USED IN NEW STEP
        CALL DAUX(T,N,X,H)
C       REARRANGE STORAGE OF PREVIOUS DERIVATIVES
        DO 16 I=1,N
        N8I=I+N8
        N7I=I+N7
        N6I=I+N6
        N2I=I+N2
C       Y PRIME (N-2) GOES TO (N-3)
        T(N8I)=T(N7I)
C       Y PRIME (N-1) GOES TO (N-2)
        T(N7I)=T(N6I)
C       Y PRIME (N) GOES TO (N-1)
     16 T(N6I)=T(N2I)
     17 RETURN
        END
```

5

Sukhanov's Variable Initial Value Method for Boundary Value Problems

In 1983, Sukhanov presented a new method for transforming a nonlinear two-point boundary value problem into an initial value problem (Reference 1). Sukhanov's approach involves only the solution of ordinary differential equations and not partial differential equations as in other initial value methods for nonlinear two-point boundary value problems (References 2, 3, and 8). In this chapter, Sukhanov's method is applied to second-order and fourth-order nonlinear two-point boundary value problems.

Consider the two-point boundary value problem consisting of the second-order system of nonlinear ordinary differential equations and known boundary conditions (References 4–6)

$$\dot{u} = F(u, v), \qquad u(0) = u_0, \tag{5.1}$$

$$\dot{v} = G(u, v), \qquad 0 \leq t \leq T, \qquad v(T) = c. \tag{5.2}$$

Boundary conditions are given at both ends of the interval $(0, T)$, one at the beginning and one at the opposite end. The value of the dependent variable $u(t)$ for a particular value of the independent variable t also depends on the interval length T and the boundary condition c. Let the unknown boundary conditions be represented by

$$r(T) = u(T) = \text{missing terminal value of } u, \tag{5.3}$$

$$s(T) = v(0) = \text{missing initial value of } v. \tag{5.4}$$

The two-point boundary value problem is transformed into an initial value problem by solving for the missing initial condition, $s(T)$. Sukhanov refers to this as a variable initial value problem.

Consider also the two-point boundary value problem consisting of the fourth-order system of nonlinear differential equations and known boundary conditions (Reference 7)

$$\dot{x}(t) = F(t, x, y, u, v), \qquad x(0) = c_1, \tag{5.5}$$

$$\dot{y}(t) = G(t, x, y, u, v), \qquad y(0) = c_2, \tag{5.6}$$

$$\dot{u}(t) = P(t, x, y, u, v), \qquad u(T) = c_3, \tag{5.7}$$

$$\dot{v}(t) = Q(t, x, y, u, v), \qquad v(T) = c_4. \tag{5.8}$$

Two of the boundary conditions are assumed to be given at the initial interval length, $t = 0$, and the other two boundary conditions are assumed to be given at the terminal interval length, $t = T$. The unknown boundary conditions are a function of the terminal length T, and may be represented by the equations

$$r(T) = x(T), \tag{5.9}$$

$$s(T) = y(T), \tag{5.10}$$

$$m(T) = u(0), \tag{5.11}$$

$$n(T) = v(0). \tag{5.12}$$

The two-point boundary value problem is transformed into an initial value problem by solving for the unknown initial conditions, $m(T)$ and $n(T)$.

Two possible methods of solution of the initial value equations are described in Section 5.1. These methods are (a) Euler's method and (b) the Runge–Kutta integration method. In Section 5.1.1, Euler's method is used to predict the unknown boundary condition of the initial value equation for the second-order system, and the Newton–Raphson method is used to correct the prediction (Reference 4). In Section 5.1.2, the unknown boundary conditions are obtained using a Runge–Kutta integration procedure for both the second- and fourth-order systems (References 5–7). The derivatives for the Runge–Kutta integration method are evaluated automatically using the automatic derivative evaluation method. The automatic derivative evaluation subroutines, as applied to Sukhanov's method, are described in Section 5.2. Numerical examples are given in Section 5.3, and a program listing is given in Section 5.4.

5.1. Sukhanov's Initial Value Equations

For the second-order system given by equations (5.1) and (5.2), Sukhanov first considers the auxiliary equations

$$\dot{U} = F(U, V), \qquad U(a, 0) = u_0, \tag{5.13}$$

$$\dot{V} = G(U, V), \qquad V(a, 0) = a, \tag{5.14}$$

$$\dot{U}_a = F_U U_a + F_V V_a, \qquad U_a(a, 0) = 0, \tag{5.15}$$

$$\dot{V}_a = G_U U_a + G_V V_a, \qquad V_a(a, 0) = 1, \tag{5.16}$$

where

$$F_U = \partial F/\partial U, \qquad U_a = \partial U/\partial a, \qquad \text{etc.} \tag{5.17}$$

The dot represents differentiation of U, V, U_a, and V_a with respect to t. Equations (5.15) and (5.16) are obtained by simply differentiating equations (5.13) and (5.14) with respect to a.

The values of $s(T)$ and $r(T)$ are connected by the relations

$$U\big(s(T), T\big) = r(T), \tag{5.18}$$

$$V\big(s(T), T\big) = c. \tag{5.19}$$

Differentiating equations (5.18) and (5.19) with respect to T yields

$$U_a(s, T)s_T + U_T(s, T) = r_T, \tag{5.20}$$

$$V_a(s, T)s_T + V_T(s, T) = 0. \tag{5.21}$$

Substituting from equations (5.13) and (5.14) into equations (5.20) and (5.21) yields

$$U_a(s, T)s_T + F(r, c) = r_T, \tag{5.22}$$

$$V_a(s, T)s_T + G(r, c) = 0. \tag{5.23}$$

5.1.1. Euler's Method

When the interval length $T = 0$, we know

$$s(0) = c, \qquad r(0) = u_0. \tag{5.24}$$

Assume now that $s(T)$ and $r(T)$ are known, $T > 0$, and we wish to advance the solution of equations (5.22) and (5.23) to $T + \Delta T$. Equations (5.13)–(5.16) are integrated from $t = 0$ to $t = T$ with $a = s(T)$. This yields the value of $V_a(s(T), T)$ and enables us to use equation (5.23) to evaluate $s_T(T)$. A preliminary estimate of $s(T + \Delta T)$, a_1, is then obtained from

$$a_1 = s(T) + s_T(T)\,\Delta T, \tag{5.25}$$

where

$$s_T = -G(r, c)/V_a(s, T) \tag{5.26}$$

Using this value, $a = a_1$, we again integrate equations (5.13)–(5.16) for $0 \leq t \leq T + \Delta T$. Then we use the Newton–Raphson correction to obtain an improved estimate of $a = s(T + \Delta T)$,

$$a_2 = a_1 - [V(a_1, T + \Delta T) - c]/V_a(a_1, T + \Delta T). \tag{5.27}$$

The numerical solution is then obtained utilizing equations (5.13)–(5.16) and (5.24)–(5.27) as follows. Let the terminal length T equal zero and ΔT equal, say, 0.1. Obtain a_1 from equation (5.25) and integrate equations (5.13)–(5.16) from $t = 0$ to $t = T + \Delta T$ using a fourth-order Runge–Kutta method with grid intervals Δt equal to, say, 0.01. Obtain a better estimate of the parameter a using equation (5.27) and again integrate equations (5.13)–(5.16) from $t = 0$ to $t = T = 0.1$. Now increment T by ΔT again and predict a_1 from equation (5.25) for $T = 0.1$ and $\Delta T = 0.1$. Integrate equations (5.13)–(5.16) from $t = 0$ to $t = T + \Delta T = 0.2$, obtain a better estimate of a using equation (5.27), and integrate equations (5.13)–(5.16) from $t = 0$ to $t = T = 0.2$. Then repeat the process for $T + \Delta T = 0.3$, etc. until the desired terminal length is reached. At each step several applications of equation (5.27) followed by integration of equations (5.13)–(5.16) could be made.

Sukhanov's method estimates the unknown initial condition, $V(0) = a$, over successively larger length intervals, by finding the boundary condition for the interval length T and then incrementing by ΔT, instead of estimating the parameter a over the entire long interval.

5.1.2. Runge–Kutta Integration Method

The solution can also be obtained using the Runge–Kutta integration method in place of Euler's method. This method will be described for both the second-order and fourth-order two-point boundary value problems.

5.1.2.1. *Second-Order System*

Solving equations (5.22) and (5.23) for r_T and s_T, we obtain

$$r_T = F(r, c) - [G(r, c)/V_a(s, T)]U_a(s, T), \qquad r(0) = u_0, \tag{5.28}$$

$$s_T = -G(r, c)/V_a(s, T), \qquad s(0) = c. \tag{5.29}$$

Sukhanov's equations for the Runge–Kutta integration method are equations (5.13)–(5.16) and (5.28)–(5.29).

The automatic derivative evaluation program is used with the Runge–Kutta integration method. Equations (5.28) and (5.29) are integrated in the main program using a fourth-order Runge–Kutta method with grid intervals $\Delta T = h$. Equations (5.13)–(5.16) are integrated in a subroutine four times for each step in the main program. For example, let

$$r_T = g(T, r, s), \tag{5.30}$$

$$s_T - f(T, r, s). \tag{5.31}$$

The Runge–Kutta integration equations are given in Table 5.1, where k_1, k_2, k_3, and k_4 are the results of the intermediate calculations for s_T. Similar equations for m_1, m_2, m_3, and m_4 apply to the integration of r_T.

In order to evaluate the Runge–Kutta equations, $V_a(s_i, T_i)$ and $U_a(s_i, T_i)$ must be evaluated [see equations (5.28) and (5.29)]. For example, to evaluate k_1 in Table 5.1, equations (5.13)–(5.16) must be integrated from $t = 0$ to $t = t_f = t_i$ with initial condition $V(a, 0) = u = s_i$, in order to obtain $V_a(s_i, T_i)$. To evaluate k_2 in Table 5.1, equations (5.13)–(5.16) must be integrated again, but this time from $t = 0$ to $t - t_f = t_i + h/2$

Table 5.1. Runge–Kutta Integration Equations

Main program Runge–Kutta integration equations	To evaluate the Runge–Kutta equations on the left, use subroutine to integrate equations (5.13)–(5.16) from $t = 0$ to $t - t_f$ with initial condition a	
$k_1 = f(T_i, r_i, s_i)$	$t_f = t_i$,	$a = s_i$
$k_2 = f(T_i + h/2, r_i + m_1h/2, s_i + k_1h/2)$	$t_f = t_i + h/2$,	$a = s_i + k_1h/2$
$k_3 = f(T_i + h/2, r_i + m_2h/2, s_i + k_2h/2)$	$t_f = t_i + h/2$,	$a = s_i + k_2h/2$
$k_4 = f(T_i + h, r_i + m_3h, s_i + k_3h)$	$t_f = t_i + h$,	$a = s_i + k_3h$
$s_{i+1} = s_i + (h/6)[k_1 + 2k_2 + 2k_3 + k_4]$		

with initial condition $V(a, 0) = a = s_i + k_1h/2$. In all, equations (5.13)–(5.16) must be integrated four times from $t = 0$ to $t = t_f$ for each grid step, $\Delta T = h$, or Runge–Kutta cycle in the main program given in Table 5.1 for the integration of equations (5.28) and (5.29). Equations (5.13)–(5.16) are integrated using a fourth-order Runge–Kutta method with grid intervals $\Delta t = h/2$. This is one-half the grid interval size, $\Delta T = h$, for the integration of equations (5.28) and (5.29), and is required to evaluate the Runge–Kutta equations at the intermediate steps in Table 5.1. The equations are initialized with $V(a, T) = V(a, 0) = c$.

5.1.2.2. Fourth-Order System

For the fourth-order two-point boundary value problem given by equations (5.5)–(5.8), Sukhanov considers the auxiliary equations

$$\dot{X} = F(t, X, Y, U, V) \qquad X(0) = c_1, \tag{5.32}$$

$$\dot{Y} = G(t, X, Y, U, V), \qquad Y(0) = c_2, \tag{5.33}$$

$$\dot{U} = P(t, X, Y, U, V), \qquad U(0) = a, \tag{5.34}$$

$$\dot{V} = Q(t, X, Y, U, V), \qquad V(0) = b, \tag{5.35}$$

where $X = X(t, a, b)$, $Y = Y(t, a, b)$, $U = U(t, a, b)$, and $V = V(t, a, b)$ are the solutions of equations (5.5)–(5.8) with unknown initial conditions a and b.

Differentiating equations (5.32)–(5.35) with respect to a and b we obtain

$$\dot{X}_a = F_X X_a + F_Y Y_a + F_U U_a + F_V V_a, \qquad X_a(0) = 0, \tag{5.36}$$

$$\dot{Y}_a = G_X X_a + G_Y Y_a + G_U U_a + G_V V_a, \qquad Y_a(0) = 0, \tag{5.37}$$

$$\dot{U}_a = P_X X_a + P_Y Y_a + P_U U_a + P_V V_a, \qquad U_a(0) = 1, \tag{5.38}$$

$$\dot{V}_a = Q_X X_a + Q_Y Y_a + Q_U U_a + Q_V V_a, \qquad V_a(0) = 0, \tag{5.39}$$

$$\dot{X}_b = F_X X_b + F_Y Y_b + F_U U_b + F_V V_b, \qquad X_b(0) = 0, \tag{5.40}$$

$$\dot{Y}_b = G_X X_b + G_Y Y_b + G_U U_b + G_V V_b, \qquad Y_b(0) = 0, \tag{5.41}$$

$$\dot{U}_b = P_X X_b + P_Y Y_b + P_U U_b + P_V V_b, \qquad U_b(0) = 0, \tag{5.42}$$

$$\dot{V}_b = Q_X X_b + Q_Y Y_b + Q_U U_b + Q_V V_b, \qquad V_b(0) = 1, \tag{5.43}$$

where the dot represents differentiation of $X_a, Y_a, \ldots, V_b$ with respect to t and

$$F_X = \partial F/\partial X, \qquad F_Y = \partial F/\partial Y, \qquad \text{etc.} \tag{5.44}$$

Equations (5.36)–(5.39) can be expressed in the form

$$\begin{bmatrix} \dot{X}_a \\ \dot{Y}_a \\ \dot{U}_a \\ \dot{V}_a \end{bmatrix} = \begin{bmatrix} F_X & F_Y & F_U & F_V \\ G_X & G_Y & G_U & G_V \\ P_X & P_Y & P_U & P_V \\ Q_X & Q_Y & Q_U & Q_V \end{bmatrix} \begin{bmatrix} X_a \\ Y_a \\ P_a \\ Q_a \end{bmatrix} \tag{5.45}$$

or

$$\dot{\mathbf{X}}_a = A\mathbf{X}_a. \tag{5.46}$$

Similarly,

$$\dot{\mathbf{X}}_b = AX_b. \tag{5.47}$$

We can now write the four relations connecting $m(T)$, $n(T)$, $r(T)$, $s(T)$, and T

$$X\big(T, m(T), n(T)\big) = r(T), \tag{5.48}$$

$$Y\big(T, m(T), n(T)\big) = s(T), \tag{5.49}$$

$$U\big(T, m(T), n(T)\big) = c_3, \tag{5.50}$$

$$V\big(T, m(T), n(T)\big) = c_4. \tag{5.51}$$

Differentiating equations (5.48)–(5.51) with respect to T yields

$$X_T + X_a m_T + X_b n_T = r_T, \tag{5.52}$$

$$Y_T + Y_a m_T + Y_b n_T = s_T, \tag{5.53}$$

$$U_T + U_a m_T + U_b n_T = 0, \tag{5.54}$$

$$V_T + V_a m_T + V_b n_T = 0. \tag{5.55}$$

Substituting from equations (5.32)–(5.35) into equations (5.52)–(5.55), noting that $X_T = F(T, r, s, c_3, c_4)$, $Y_T = G(T, r, s, c_3, c_4)$, etc. and rearranging yields

$$r_T = F(T, r, s, c_3, c_4) + X_a m_T + X_b n_T, \qquad r(0) = c_1, \tag{5.56}$$

$$s_T = G(T, r, s, c_3, c_4) + Y_a m_T + Y_b n_T, \qquad s(0) = c_2, \tag{5.57}$$

$$m_T = [Q(T, r, s, c_3, c_4)U_b - P(T, r, s, c_3, c_4)V_b]/D, \qquad m(0) = c_3, \tag{5.58}$$

$$n_T = [P(T, r, s, c_3, c_4)U_b - Q(T, r, s, c_3, c_4)U_a]/D, \qquad n(0) = c_4, \tag{5.59}$$

where

$$D = U_a V_b - U_b V_a \tag{5.60}$$

Sukhanov's equations are equations (5.32)–(5.43) and (5.56)–(5.59).

The unknown boundary conditions are obtained by integrating equations (5.56)–(5.59). This is done using the Runge–Kutta integration procedure as follows. Assume that $r(T)$, $s(T)$, $m(T)$, and $n(T)$ are known for a given value of T. Then to step forward to $T + \Delta T$, the values of $X_a\big(T, m(T), n(T)\big)$, $X_b\big(T, m(T), n(T)\big)$, $Y_a\big(T, m(T), n(T)\big)$, etc. are required. These values are obtained by integrating equations (5.32)–(5.43) from $t = 0$ to $t = T$ with the initial conditions $a = m(T)$ and $b = n(T)$. The derivatives in the A matrix of equation (5.46), [where equations (5.46) and (5.47) are equivalent to equations (5.36)–(5.43)], are computed automatically using the automatic calculation of derivatives method.

Equations (5.56)–(5.59) are integrated in the main program using a fourth-order Runge–Kutta method with grid intervals $\Delta T = h$. Equations (5.32)–(5.43) are integrated in a subroutine four times for each step in the main program. In order to explain the procedure, assume that m_T is a function of only T and m:

$$m_T = f(T, m). \tag{5.61}$$

The Runge–Kutta integration equations for m_T are given in Table 5.2.

In order to evaluate the main program Runge–Kutta equations, equations (5.32)–(5.43) must be evaluated, such as $U_a(T_i, m_i, n_i)$ at the terminal boundary T_i. For example to evaluate k_1 in Table 5.2, equations (5.32)–(5.43) must be integrated from $t = 0$ to $t = t_f = t_i$ with initial condition $U(0) = a = m_i$, in order to obtain $U_a(T_i, m_i, n_i)$. To evaluate k_2 in Table 5.2, equations (5.32)–(5.43) must be integrated again, but this time from $t = 0$ to $t = t_f = t_i + h/2$ with initial condition $U(0) = a = m_i + k_1h/2$. In all, equations (5.32)–(5.43) must be integrated four times from $t = 0$ to

Table 5.2. Runge–Kutta Integration Equations for $m_T = f(T, m)$

Main program Runge–Kutta integration equations	To evaluate the Runge–Kutta equations on the left, use subroutine to integrate equations (5.32)–(5.43) from $t = 0$ to $t = t_f$ with initial condition a	
$k_1 = f(T_i, m_i)$	$t_f = t_i,$	$a = m_i$
$k_2 = f(T_i + h/2, m_i + k_1h/2)$	$t_f = t_i + h/2,$	$a = m_i + k_1h/2$
$k_3 = f(T_i + h/2, m_i + k_2h/2)$	$t_f = t_i + h/2,$	$a = m_i + k_2h/2$
$k_4 = f(T_i + h, m_i + k_3h)$	$t_f = t_i + h,$	$a = m_i + k_3h$
$m_{i+1} = m_i + (h/6)[k_1 + 2k_2 + 2k_3 + k_4]$		

$t = t_f$ for each grid step, $\Delta T = h$, or Runge–Kutta cycle in the main program given in Table 5.2 for the integration of equations (5.56)–(5.59).

Note that m_T is actually a function of r, s, m, and n, but was assumed to be a function of only m in Table 5.2 in order to explain the Runge–Kutta procedure.

Equations (5.32)–(5.43) are integrated using a fourth-order Runge–Kutta method with grid intervals $\Delta t = h/2$. This is one-half the grid interval size, $\Delta T = h$, for the integration of equations (5.56)–(5.59), and is required to evaluate the Runge–Kutta equations at the intermediate steps in Table 5.2.

5.2. Automatic Derivative Evaluation

The automatic derivative evaluation method was used only for the Runge–Kutta integration method. The subroutines for the automatic derivative evaluation are basically the same for both the second-order and fourth-order two-point boundary value problems. Only the dimensions of the vectors and the definitions of the vector components change.

5.2.1. Second-Order System

The partial derivatives of the functions F and G in equations (5.13)–(5.16) and (5.28)–(5.29) are computed automatically, so that once coded, the same equations apply for any F and G. For example, expressing F as a function of all the variables

$$\begin{aligned} F &= F(U, V) \\ &= F(t, U, V, U_a, V_a, s, r), \end{aligned} \tag{5.62}$$

the partial derivatives are $\partial F/\partial t$, $\partial F/\partial U$, etc.

The vector $F(I)$ in the table method consists of eight components as shown in Table 5.3. The first component corresponds to $F(t, U, V, U_a, V_a, s, r)$ and the next seven components correspond to the partial derivatives. The definitions of the vector components are given in Table 5.4.

In order to compute the derivatives automatically, an eight-component vector must be computed for each of the seven variables, t, U, V, U_a, V_a, s, r and for each of the functions of the seven variables, such as the sum, the product, the square root, the quotient, etc. The subroutines required to compute these vectors are as follows: (1) linear, (2) constant,

Table 5.3. Number of Components of the Vectors **F** and **G** for the Second-Order Nonlinear Two-Point Boundary Value Problem

Function	Number of components
F or G	1
First derivatives	7

(3) add, (4) multiplication, (5) division, and (6) function. Each of these subroutines is discussed in more detail below. The computation and storage of the derivatives in the above manner is called the table method. The equations to be integrated are represented in the main integration program by the two-component vectors **TZ** and **DTZ**, for s, r and s_T, r_T, respectively, and in the integration subroutine by the four-component vectors **Z** and **DZ** for U, V, U_a, V_a and their derivatives with respect to t, respectively.

5.2.1.1. *Subroutine Linear*

In subroutine LIN the variables t, U, V, U_a, V_a, s, and r are represented by $X(1)$, $X(2)$, ..., $X(7)$, respectively, and the corresponding vectors are represented by the submatrices **A(I, J)**, i.e., the vectors **A(1, J)**, **A(2, J)**, ..., **A(7, J)**, where I designates the vector corresponding to the variable I given in Table 5.4, $I = 1, \ldots, 7$; and J designates the component of the vector corresponding to the value of the function or its derivatives J given in Table 5.4, $J = 1, \ldots, 8$. The first derivatives of the variables are equal to unity, while all the other derivatives are equal to zero. Thus

$$A(1, 1) = t, \qquad A(2, 1) = U, \qquad \ldots, \qquad A(7, 1) = r, \tag{5.63}$$

$$A(1, 2) = 1, \qquad A(2, 3) = 1, \qquad \ldots, \qquad A(7, 8) = 1, \tag{5.64}$$

and all the other components of **A(I, J)** are equal to zero. The definitions of the FORTRAN vectors corresponding to the variables and the rows of the matrix A are given in Table 5.5.

Equations (5.13)–(5.16) and (5.28)–(5.29) show that only the partial derivatives of F and G with respect to U and V are required for Sukhanov's method. The quantities such as U_a, V_a, r, and s are variables obtained by integrating equations (5.13)–(5.16) and (5.28)–(5.29). Thus vectors for only the first three variables in Table 5.5 need be defined. Usually, however,

Table 5.4. Definitions of the Vector Components for the Table Method and the Second-Order System

Vector component I or J	Variable $X(I)$	Component definitions of vectors corresponding to a variable or a function such as **A**(2, **J**) or **F**(**J**), respectively[a]	Examples	
			Vector **A**(2, **J**) corresponding to a variable	Vector **F**(**J**) corresponding to a function
1	t	z	U	$F(U, V)$
2	U	z_t	0	F_t
3	V	z_U	1	F_U
4	U_a	z_V	0	F_V
5	V_a	z_{U_a}	0	F_{U_a}
6	s	z_{V_a}	0	F_{V_a}
7	r	z_s	0	F_s
8		z_r	0	F_r

[a] $z_i = \partial z/\partial i$, where z is a scalar variable equal to t, U, V, U_a, V_a, s, or r or a function of the variables, such as F or G.

Table 5.5. FORTRAN Representation of the Variables for the Second-Order Nonlinear Two-Point Boundary Value Problem

Variable	FORTRAN program		
	Variable	Vector	Submatrix[a]
t	$X(1)$	**T**	$A(1, J)$
U	$X(2)$	**X1**	$A(2, J)$
V	$X(3)$	**X2**	$A(3, J)$
U_a	$X(4)$	**X3**	$A(4, J)$
V_a	$X(5)$	**X4**	$A(5, J)$
s	$X(6)$	**X5**	$A(6, J)$
r	$X(7)$	**X6**	$A(7, J)$

[a] The vectors **T**, **X1**, ..., **X6** are set equal to the submatrices, $A(I, J)$, $I = 1, \ldots, 7$ because the former are easier to manipulate in subroutine INPUT. Note that for Sukhanov's method partial derivatives are not required with respect to all the variables. Thus only the first three vectors need be defined.

the table method would be set up to compute the partial derivatives with respect to all the variables.

5.2.1.2. *Subroutine Constant*

In subroutine CONST, a constant term, such as a constant coefficient multiplier in the system equation, is represented by the scalar CON and its corresponding vector is represented by **D**. The first component, $D(1) = \text{CON}$, and all the derivatives, $D(2)$, $D(3)$, ..., $D(8)$, are equal to zero.

5.2.1.3. *Subroutine Add*

In subroutine ADD, two functions, D and E, and their derivatives are added:

$$G = D + E, \qquad \text{functions,} \tag{5.65}$$

$$G_i = D_i + E_i, \qquad \text{first derivatives,} \tag{5.66}$$

where

$$G_i = \partial G / \partial i, \tag{5.67}$$

$i = 1, 2, \ldots, 7$ corresponding to t, U, V, U_a, V_a, s, or r, respectively. Using the table method the two vectors, **D(I)** and **E(I)** are simply added

$$\mathbf{G(I)} = \mathbf{D(I)} + \mathbf{E(I)}, \qquad I = 1, 2, \ldots, 8. \tag{5.68}$$

A subroutine for subtraction can be obtained by simply setting the vector **E** equal to $-\mathbf{E}$. The latter is done by multiplying **E** by the constant, -1.

5.2.1.4. *Subroutine Multiplication*

In subroutine MULT, two functions, A and B are multiplied:

$$E = AB. \tag{5.69}$$

The first derivatives are

$$E_i = A_i B + A B_i. \tag{5.70}$$

5.2.1.5. *Subroutine Division*

In subroutine DIV, the derivatives of the quotient, B/U, are obtained where the numerator is the function B and the denominator is the function U. The derivatives of the reciprocal, $R = 1/U$ are calculated and then subroutine MULT is called to obtain the function $F1 = B*R$ and its derivatives.

Consider the function

$$R = f(u). \tag{5.71}$$

The first derivatives are

$$R_i - f'(u)u_i, \tag{5.72}$$

where

$$f(u) = u^{-1}, \tag{5.73}$$

$$f'(u) = -u^{-2}. \tag{5.74}$$

Equations (5.73) and (5.74) are evaluated in subroutine DIV. Equations (5.71) and (5.72) are then evaluated by calling subroutine DER. The method of calculating the eight-component vector of the function R and its derivatives in DER is similar to the method used in MULT.

5.2.1.6. *Subroutine Function*

In addition to the arithmetic operations, add, subtract, multiply, and divide described above, certain special functions such as the square root can be used as primitives in building up other functions. Subroutine DER can be used to obtain the derivatives of any function of a single variable. For example to obtain the square root, subroutine SR evaluates the equations

$$f(u) = u^{1/2}, \tag{5.75}$$

$$f'(u) = \tfrac{1}{2}u^{-1/2}. \tag{5.76}$$

Subroutine DER is then called to evaluate equations (5.71) and (5.72), where $f(u)$ and its derivatives are defined as above.

5.2.2. Fourth-Order System

The partial derivatives of the functions F, G, P, and Q in equations (5.32)–(5.43) and (5.56)–(5.59) are computed automatically, so that once coded, the same equations apply for any F, G, P, and Q. For example, expressing F as a function of all the variables

$$\begin{aligned} F &= F(t, X, Y, U, V) \\ &= F(t, X, Y, U, V, X_a, Y_a, U_a, V_a, X_b, Y_b, U_b, V_b, r, s, m, n) \end{aligned} \tag{5.77}$$

the partial derivatives are $\partial F/\partial t$, $\partial F/\partial X$, etc.

The vector **F(J)** in the table method consists of 18 components as shown in Table 5.6. The first component corresponds to the function F and the next 17 components correspond to its first partial derivatives. The definitions of the vector components are given in Table 5.7.

In order to compute the derivatives automatically, an 18-component vector must be computed for each of the 17 variables, t, X, Y, U, V, X_a, Y_a, U_a, V_a, X_b, Y_b, U_b, V_b, r, s, m, n and for each of the functions of the 17 variables, such as the sum, the product, the square root, the quotient, etc. As discussed above the subroutines required to compute these vectors are (1) linear, (2) constant, (3) add, (4) multiplication, (5) division, and

Table 5.6. Number of Components of the Vectors **F**, **G**, **P**, and **Q** for the Fourth-Order Nonlinear Two-Point Boundary Value Problem

Function	Number of components
F, G, P, or Q	1
First derivatives	17

Table 5.7. Definitions of the Vector Components for the Table Method and the Fourth-Order System

Vector component I or J	Variable $X(I)$	Component definitions of vectors corresponding to a variable or a function such as **X2(J)** or **F(J)**, respectively[a]	Example: vector **X2(J)** corresponding to a variable
1	t	z	X
2	X	z_t	0
3	Y	z_X	1
4	U	z_Y	0
5	V	z_U	0
6	X_a	z_V	0
7	Y_a	z_{X_a}	0
8	U_a	z_{Y_a}	0
9	V_a	z_{U_a}	0
10	X_b	z_{V_a}	0
11	Y_b	z_{X_b}	0
12	U_b	z_{Y_b}	0
13	V_b	z_{U_b}	0
14	r	z_{V_b}	0
15	s	z_r	0
16	m	z_s	0
17	n	z_m	0
18		z_n	0

[a] $z_i = \partial z/\partial i$, where z is a scalar variable equal to t, X, Y, U, V, X_a, Y_a, U_a, V_a, X_b, Y_b, U_b, V_b, r, s, m, or n or a function of the variables, such as F, G, P, or Q. Note that $X(I)$ in column 2 is a vector whereas X is a scalar.

(6) function. Subroutine linear is discussed in detail below. The other subroutines are the same as for the second-order system discussed above, except that the dimensions of the derivative vectors are 17 instead of 8. The equations to be integrated are represented in the main integration program by the four-component vectors **TZ** and **DTZ** for r, s, m, n and r_T, s_T, m_T, n_T, respectively, and in the integration subroutine by the 12-component vectors **Z** and **DZ** for X, Y, U, V, X_a, Y_a, U_a, V_a, X_b, Y_b, U_b, V_b and their derivatives with respect to t, respectively.

In subroutine LIN the variables, t, X, Y, U, V, X_a, Y_a, U_a, V_a, X_b, Y_b, U_b, V_b, r, s, m, and n are represented by $X(1)$, $X(2)$, . . . , $X(17)$, respectively, and the corresponding variables and their derivatives are represented by the rows of the 17×18 matrix A. The vectors corresponding to the variables and the rows of matrix A are represented in subroutine INPUT

Table 5.8. FORTRAN Representation of the Variables for the Fourth-Order Nonlinear Two-Point Boundary Value Problem

Variable	FORTRAN program	
	Variable	Vector[a]
t	$X(1)$	**T**
X	$X(2)$	**X1**
Y	$X(3)$	**X2**
U	$X(4)$	**X3**
V	$X(5)$	**X4**
X_a	$X(6)$	**X5**
Y_a	$X(7)$	**X6**
U_a	$X(8)$	**X7**
V_a	$X(9)$	**X8**
X_b	$X(10)$	**X9**
Y_b	$X(11)$	**X10**
U_b	$X(12)$	**X11**
V_b	$X(13)$	**X12**
r	$X(14)$	**X13**
s	$X(15)$	**X14**
m	$X(16)$	**X15**
n	$X(17)$	**X16**

[a] Note that for Sukhanov's method partial derivatives are not required with respect to all the variables. Thus only the first five vectors need be defined above.

by **T**, **X**1, **X**2, ..., **X**16 as shown in Table 5.8. The first derivatives of the variables are equal to unity, while all the other derivatives are equal to zero:

$$T(1) = t, \qquad X1(1) = X, \qquad X2(1) = Y, \qquad \ldots, \qquad X16(1) = n, \tag{5.78}$$

$$T(2) = 1, \qquad X1(3) = 1, \qquad X2(4) = 1, \qquad \ldots, \qquad X16(18) = 1. \tag{5.79}$$

All other components of **T**, **X**1, **X**2, ..., **X**16 are equal to zero.

Equations (5.32)–(5.43) and (5.56)–(5.59) show that only the partial derivatives of F, G, P, and Q with respect to X, Y, U, and V are required for Sukhanov's method. The quantities such as X_a, Y_a, ..., r, s, m, and n are variables obtained by integrating equations (5.32)–(5.43) and (5.56)–(5.59). Thus vectors for only the first five variables in Table 5.8 need be defined, i.e., A is a 5×18 matrix. Usually the table method would be set up to compute the partial derivatives with respect to all the variables and A would be a 17×18 matrix.

5.3. Examples Using Sukhanov's Method

Two examples using Sukhanov's method for two-point boundary value problems are given in this section. A second-order system is evaluated using both Euler's method and the Runge–Kutta integration method. A fourth-order system is evaluated using only the latter method. The automatic derivative evaluation method is used with the Runge–Kutta integration procedure.

5.3.1. Second-Order System

Consider the two-point boundary value problem

$$\dot{u} = v, \qquad u(0) = 0, \tag{5.80}$$

$$\dot{v} = e^{u}, \qquad v(0.5) = 0 = c. \tag{5.81}$$

Using Euler's method described in Section 5.1.1, equations (5.13)–(5.16) become

$$\dot{U} = V, \qquad U(0) - 0, \tag{5.82}$$

$$\dot{V} = e^{U}, \qquad V(0) = a, \tag{5.83}$$

$$\dot{U}_a = V_a, \qquad U_a(0) = 0, \tag{5.84}$$

$$\dot{V}_a = e^{U} U_a, \qquad V_a(0) = 1. \tag{5.85}$$

At $T = 0$, the initial a is

$$a_1 = s(T) - \frac{G(r, c)}{V_a(s, T)} \Delta T = - \frac{e^{U(0)}}{V_a(0)} \Delta T = (-1) \Delta T. \tag{5.86}$$

The numerical results are given in Table 5.9. Equation (5.25) is used to predict a for $T = 0$ and $\Delta T = 0.1$. Equations (5.82)–(5.85) are integrated from $t = 0$ to $t = T + \Delta T = 0.1$ using a fourth-order Runge–Kutta method with grid intervals, $\Delta t = 0.01$. Equation (5.27) is used to obtain a better estimate of a, and equations (5.82)–(5.85) are integrated again from $t = 0$ to $t + T = 0.2$. The length interval is then incremented by $\Delta T = 0.1$ again and the above process is repeated until the desired terminal length, $T = 0.5$ is reached.

The analytical solution for the above two-point boundary value problem is (Reference 1)

$$u(t) = \ln \frac{C_1^{2}}{1 + \cos C_1(t - C_2)}, \tag{5.87}$$

Table 5.9. Numerical Results for the Second-Order System Example Using Euler's Method[a]

Integration length interval (see note a below)	T	ΔT	a	t	U	V
(1)	0.0	0.1	−0.1	0.0	0.0	−0.1
				0.1	$-0.5012486\,E-2$	$-0.3329982\,E-3$
(2)	0.1	0.0	$-0.9966865\,E-1$	0.0	0.0	$-0.9966865\,E-1$
				0.1	$-0.4979296\,E-2$	$-0.3492460\,E-9$
(1)	0.1	0.1	−0.1986786	0.0	0.0	−0.1986786
				0.1	$-0.1489670\,E-1$	$-0.9950169\,E-1$
				0.2	$-0.1993308\,E-1$	$-0.1308357\,E-2$
(2)	0.2	0.0	−0.1973955	0.0	0.0	−0.1973955
⋮				0.1	$-0.1476818\,E-1$	$-0.9821230\,E-1$
				0.2	$-0.1967478\,E-1$	$0.3143214\,E-8$
(1)	0.4	0.1	−0.4667035	0.0	0.0	−0.4667035
				0.1	$-0.4174322\,E-1$	−0.3688418
				0.2	$-0.7388651\,E-1$	−0.2745294
				0.3	$-0.9673425\,E-1$	−0.1827748
				0.4	−0.1104969	$-0.9268457\,E-1$
				0.5	−0.1152989	$-0.3426697\,E-2$
(2)	0.5	0.0	−0.4636324	0.0	0.0	−0.4636324
				0.1	$-0.4143561\,E-1$	−0.3657557
				0.2	$-0.7326835\,E-1$	−0.2713998
				0.3	$-0.9579980\,E-1$	−0.1795741
				0.4	−0.1092377	$-0.8938506\,E-1$
				0.5	−0.1137036	$0.1919689\,E-6$

[a] Note: (1) Integration from $t = 0$ to $T + \Delta T$. Initial condition, $V(0) = a = a_1$. (2) Integration from $t = 0$ to T. Initial condition, $V(0) = a = a_2$.

where

$$C_1^2 = 1.785044819891043, \tag{5.88}$$

$$C_2 = 0.5. \tag{5.89}$$

At $t = 0.5$, this gives

$$u(0.5) = -0.11370365646. \tag{5.90}$$

The derivative of equation (5.87) is

$$v(t) = \dot{u}(t) = C_1 \tan \frac{C_1(t - C_2)}{2}. \tag{5.91}$$

At $t = 0$, this gives

$$v(0) = -0.463632592. \tag{5.92}$$

Table 5.9 shows that the numerical results at $t = 0.5$ are accurate to at least six or seven decimal places.

Numerical results were also obtained using $T = 0$ and $\Delta T = 0.5$ for the initial integration interval. Equation (5.27) was then used to obtain a better estimate of a and equations (5.82)–(5.85) were integrated from $t = 0$ to $t = T = 0.5$. The Newton–Raphson method was then repeated, i.e., equation (5.27) was utilized again and equations (5.82)–(5.85) were again integrated from $t = 0$ to $t = T = 0.5$. The terminal value of $u(t)$ this time was $u(0.5) = -0.1137037$, again accurate to approximately seven decimal places.

Using the Runge–Kutta integration method, again consider the two-point boundary value problem

$$\dot{u} = v, \qquad u(0) = 0, \tag{5.93}$$

$$\dot{v} = e^U, \qquad v(0.5) = 0 = c. \tag{5.94}$$

Equations (5.13) and (5.14) then become

$$\dot{U} = F = V, \qquad U(0) = 0, \tag{5.95}$$

$$\dot{V} = G = e^U, \qquad V(0) = a. \tag{5.96}$$

Equations (5.95) and (5.96) are entered into the automatic derivative evaluation program in subroutine INPUT. Subroutine INPUT calls the subroutines given in Table 5.10. In all cases the subroutines define the variables

Table 5.10. Subroutine INPUT for the Second-Order System Example

Subroutine INPUT	Purpose
CALL LIN(X)	Defines the vectors corresponding to the variables t, U, V, U_a, V_a, s, and r.
CALL CONST(1.,D)	Defines the vector D corresponding to the constant $= 1$.
CALL MULT(D,V,F)	Multiplies D times V to form $F = V$ and its derivatives.
CALL AEXP(U,G)	Forms the function $G = e^U$ and its derivatives.

or functions indicated and their derivatives. Additional boundary conditions are

$$U_a(0) = 0, \qquad V_a(0) = 1, \tag{5.97}$$

$$r(0) = 0, \qquad s(0) = 0. \tag{5.98}$$

At $T = 0$, the initial a is equal to $c = 0$.

No additional inputs are required from the user. Equations (5.15)–(5.16) and (5.28)–(5.29) are evaluated automatically in the program from the functions F and G given in subroutine INPUT.

The numerical results are given in Table 5.11. Equations (5.28) and (5.29) are integrated from $T = 0$ to $T = 0.5$ with grid intervals, $\Delta T = h = 0.05$. Grid intervals of $\Delta t = h/2 = 0.025$ are used for equations (5.13)–(5.16), which are integrated four times for each grid step ΔT in equations (5.28) and (5.29).

The analytical solution for the two-point boundary value problem was given above in equations (5.87)–(5.92). Table 5.11 shows that the numerical

Table 5.11. Numerical Results for the Second-Order System Example Using the Runge–Kutta Integration Method

T	$r(T)$	$s(T)$
0.0	0.0	0.0
0.1	−0.004979297	−0.09966865
0.2	−0.01967478	−0.1973955
0.3	−0.04340175	−0.2914562
0.4	−0.07514512	−0.3805026
0.5	−0.1137037	−0.4636326

results are accurate to at least seven decimal places, where

$$r(0.5) = u(0.5), \tag{5.99}$$

$$s(0.5) = v(0). \tag{5.100}$$

5.3.2. Fourth-Order System

Consider the fourth-order two-point boundary value problem

$$x^{(4)} = -ke^{x}, \qquad k = 6, \tag{5.101}$$

$$x(0) = \dot{x}(0) = 0, \tag{5.102}$$

$$\ddot{x}(1) = \dddot{x}(1) = 0. \tag{5.103}$$

Equations (5.32)–(5.35) then become

$$\dot{X} = F = Y, \qquad X(0) = 0, \tag{5.104}$$

$$\dot{Y} = G = U, \qquad Y(0) = 0, \tag{5.105}$$

$$\dot{U} = P = V, \qquad U(0) = a, \tag{5.106}$$

$$\dot{V} = Q = -ke^{X}, \qquad V(0) = b. \tag{5.107}$$

The only inputs required from the user are the functions F, G, P, and Q, the boundary conditions, and the terminal length T. The boundary conditions and the terminal length are specified in the main program. The functions F, G, P, and Q are entered in subroutine INPUT as shown in Table 5.12. In all cases the subroutines define the variables or functions indicated and their derivatives. Additional boundary conditions are

$$r(0) = 0, \qquad s(0) = 0, \tag{5.108}$$

$$m(0) = 0, \qquad n(0) = 0. \tag{5.109}$$

At $T = 0$, the initial conditions a and b are equal to zero. Equations (5.36)–(5.43) and (5.56)–(5.59) are evaluated automatically in the program from the functions F, G, P, and Q given in subroutine INPUT.

The numerical results are given in Table 5.13. Equations (5.56)–(5.59) are integrated from $T = 0$ to $T = 1$ with grid intervals, $\Delta T = h = 0.05$. Grid intervals of $\Delta t = h/2 = 0.025$ are used for equations (5.32)–(5.43), which are integrated four times for each grid step ΔT in equations (5.56)–(5.59).

Table 5.12. Subroutine INPUT for the Fourth-Order System Example

Subroutine INPUT	Purpose
CALL LIN(X)	Defines the vectors corresponding to the variables t, X, Y, U, V, X_a, Y_a, U_a, V_a, X_b, Y_b, U_b, V_b, r, s, m, and n.
CALL CONST(1.,D)	Defines the vector D corresponding to the constant one.
CALL MULT(D,Y,F)	Multiplies D times Y to form $F = Y$ and its derivatives.
CALL MULT(D,U,G)	Multiplies D times U to form $G = U$ and its derivatives.
CALL MULT(D,V,P)	Multiplies D times V to form $P = V$ and its derivatives.
CALL CONST(−6.,D1)	Defines the vector $D1$ corresponding to the constant −6.
CALL AEXP(X,E)	Forms the function $E = e^X$ and its derivatives (where X is the vector equal to X1, corresponding to the variable X and is not the same as the vector X in subroutine LIN).
CALL MULT(D1,E,Q)	Multiplies $D1$ times E to form $Q = -6e^X$ and its derivatives.

Table 5.13. Numerical Results for the Fourth-Order System Example Using the Runge–Kutta Integration Method

T	$r(T)$	$s(T)$
0.0	0.0	0.0
0.2	−0.001199057	−0.007993511
0.4	−0.01896296	−0.06318509
0.6	−0.09155688	−0.2030733
0.8	−0.2595715	−0.4302689
1.0	−0.5327300	−0.7022792
t	$X(t)$	$Y(t)$
0.0	0.0	0.0
0.2	−0.03838131	−0.3550828
0.4	−0.1320969	−0.5610157
0.6	−0.2557962	−0.6619224
0.8	−0.3925157	−0.6974136
1.0	−0.5327300	−0.7022792

$a = m(1) = -2.226207$; $b = n(1) = 4.906284$

The numerical results agree to at least six decimal places (using the DEC 20) with the results given in Reference 3 where the problem was solved using the method of quasilinearization. As another check, the problem was solved using Sukhanov's method but without the use of the automatic calculation of derivatives. Again the results were the same.

The major advantage of the implementation described in this chapter is that it provides a systematic method for obtaining the unknown initial conditions for nonlinear two-point boundary value problems with little or no effort required to compute the derivatives.

Table 5.14. Definitions of the FORTRAN Variables and Vectors for the Fourth-Order Nonlinear Two-Point Boundary Value Problem Example

	Variable	FORTRAN variable or vector	Number of components	Program line number
Main program integration:				320–600
Initial conditions	Equations (5.33)–(5.36)	**TZ**	4	278
Integration grid interval size	ΔT	**TU**	1	270
Integration limits	T	**N1*TU,(N1=20)**	1	320
State variables	See footnote *b*.	**TZ**	4	515
Derivatives to be integrated each iteration	Equations (5.33)–(5.36)	**DTZ**	4	3985–4000
Subroutine integration:				3180–3540
Initial conditions	Equations (5.9)–(5.20)	**CI**	12	140–160, 400, 440, 486, 550
Integration grid interval size	Δt	**H**	1	170
Integration limits	t_f	**L*H**	1	385, 3320
State variables	See footnote *c*.	**Z**	12	3480
Derivatives to be integrated each iteration	Equations (5.9)–(5.20)	**DZ**	12	3795–3870

[a] Additional definitions of the variables and vectors are given in Tables 5.7 and 5.8. The functions F, G, P, and Q are represented in subroutine INPUT by F1, F2, F3, and F4, respectively.
[b] $\mathbf{TZ} = (r, s, m, n)^T$, $\mathbf{DTZ} = (r_T, s_T, m_T, n_T)^T$.
[c] $\mathbf{Z} = (X, Y, U, V, X_a, Y_a, U_a, V_a, X_b, Y_b, U_b, V_b)^T$, $\mathrm{DZ} = \dot{Z}$.

5.4. Program Listing

The FORTRAN program listing is given in this section for the fourth-order nonlinear two-point boundary value problem example described in Section 5.3.2. The Runge–Kutta integration method described in Section 5.1.2 is used with the automatic calculation of the derivatives. The definitions of the FORTRAN variables and vectors used in the program are given in Table 5.14.

```
40          DIMENSION TH(4),TI(4),TJ(4),TK(4)
45          DIMENSION TY(4),TZ(4),TCI(4),M(20)
50          DIMENSION U(173),U1(173),UT(173)
55          DIMENSION X(17),HA(18),X1(101,12),A(5,18)
60          DIMENSION CI(12),Z(12),DZ(12),X0(12),P0(12)
65          DIMENSION AH(12),AI(12),AJ(12),AK(12),Y(12)
70          COMMON L,X1D,ID,L1,LV,IV,A
72          ID1=12
73          ID2=4
75          ID=ID1
80          LV=18
85          IV=17
90          DO 335 I=1,5
95          DO 340 K=1,LV
100   340    A(I,K)=0.0
105   335    CONTINUE
110          DO 345 I=1,5
120   345    A(I,I+1)=1.0
130          DO 650 I=1,ID
140   650    CI(I)=0.0
150          CI(7)=1.0
160          CI(12)=1.0
170          H=1./40.
180          X1I=0.0
190          X1L=0.1
210          X1D=X1L-X1I
215          C1=1.0
220          L=0
225          TA=0.0
250          JS=1
260          DO 355 I=1,ID
265   355    X1(1,I)=CI(I)
270          TU=1./20
275          DO 660 I=1,ID2
278   660    TZ(I)=0.0
280          TU2=TU/2.0
285          IT=ID1
290          KP=10
295          I1=1
300          IN=1
305          CALL MMULT(CI,C1,Z)
310          DO 600 I=1,20
315   600    M(I)=I
320          DO 150 N1=1,20
325          TS=TA+(N1-1)*TU
```

```
340            NL=N1
345            ID=ID2
360            CALL MMULT(TZ,C1,TY)
365            CALL TFORCE(TZ,DTZ,Z,IT)
370            CALL MMULT(DTZ,TU2,TH)
380            CALL MADD(TY,TH,TZ)
385            L=INT((TS+.1*H)/H)+1
390            CI(3)=TZ(3)
392            CI(4)=TZ(4)
395            ID=ID1
400            CALL MMULT(CI,C1,Z)
405            CALL INTEG(X,HA,X1,U,U1,Z,DZ,IN,H,KP,JS,AH,AI,AJ,AK,Y)
410            ID=ID2
415            CALL TFORCE(TZ,DTZ,Z,IT)
420            CALL MMULT(DTZ,TU2,TI)
425            CALL MADD(TY,TI,TZ)
430            CI(3)=TZ(3)
432            CI(4)=TZ(4)
435            ID=ID1
440            CALL MMULT(CI,C1,Z)
445            CALL INTEG(X,HA,X1,U,U1,Z,DZ,IN,H,KP,JS,AH,AI,AJ,AK,Y)
450            ID=ID2
455            CALL TFORCE(TZ,DTZ,Z,IT)
460            CALL MMULT(DTZ,TU,TJ)
465            CALL MADD(TY,TJ,TZ)
470            L=L+1
475            CI(3)=TZ(3)
478            CI(4)=TZ(4)
480            ID=ID1
486            CALL MMULT(CI,C1,Z)
490            CALL INTEG(X,HA,X1,U,U1,Z,DZ,IN,H,KP,JS,AH,AI,AJ,AK,Y)
495            ID=ID2
500            CALL TFORCE(TZ,DTZ,Z,IT)
510            CALL MMULT(DTZ,TU,TK)
515            CALL MMUAD(TH,TI,TJ,TK,TY,TZ)
520            M1=M(I1)
522            TS=TA+N1*TU
525            IF(N1-M1)602,605,602
530     605    TYPE*,TS,TZ(1),TZ(2)
535            I1=I1+1
540     602    CI(3)=TZ(3)
542            CI(4)=TZ(4)
545            ID=ID1
550            CALL MMULT(CI,C1,Z)
555            CALL INTEG(X,HA,X1,U,U1,Z,DZ,IN,H,KP,JS,AH,AI,AJ,AK,Y)
600     150    CONTINUE
610            TYPE 620
620     620    FORMAT(38H      X                  Z1                 Z2)
630            DO 630 I=1,41
640            TS=TA+(I-1)*H
650            TYPE*,TS,X1(I,1),X1(I,2)
660     630    CONTINUE
670            DO 750 I=1,12
680     750    TYPE*,Z(I),CI(I)
720            END
```

```
780          SUBROUTINE INPUT(X,F1,F2,F3,F4)
785          COMMON L,X1D,ID,L1,LV,IV,A
790          DIMENSION A(5,18),T(18)
795          DIMENSION X1(18),X2(18),X3(18),X4(18)
800          DIMENSION F1(LV),F2(LV),F3(LV),F4(LV)
805          DIMENSION X(IV),D(18),E1(18)
815          CALL LIN(X)
820          DO 325 I=1,LV
825          T(I)=A(1,I)
870          X1(I)=A(2,I)
875          X2(I)=A(3,I)
880          X3(I)=A(4,I)
885    325   X4(I)=A(5,I)
895          B1=1.0
900          CALL CONST(B1,D)
905          CALL MULT(D,X2,F1)
910          CALL MULT(D,X3,F2)
915          CALL MULT(D,X4,F3)
920          B2=-6.0
925          CALL CONST(B2,D)
930          CALL AEXP(X1,E1)
935          CALL MULT(D,E1,F4)
975          RETURN
980          END
985          SUBROUTINE LIN(X)
990          COMMON L,X1D,ID,L1,LV,IV,A
995          DIMENSION X(IV),A(5,18)
1000          DO 330 I=1,5
1005    330   A(I,1)=X(I)
1010          RETURN
1015          END
1090          SUBROUTINE CONST(CON,D)
1095          COMMON L,X1D,ID,L1,LV,IV
1100          DIMENSION D(LV)
1110          DO 15 I=1,LV
1120     15   D(I)=0.0
1130          D(1)=CON
1140          RETURN
1150          END
1160          SUBROUTINE MULT(A,B,E)
1165          COMMON L,X1D,ID,L1,LV,IV
1170          DIMENSION A(LV),B(LV),E(LV)
1180          E(1)=A(1)*B(1)
1190          DO 15 I=1,IV
2000     15   E(I+1)=A(I+1)*B(1)+A(1)*B(I+1)
2020          RETURN
2030          END
2040          SUBROUTINE SCOS(U,S)
2050          COMMON L,X1D,ID,L1,LV,IV
2060          DIMENSION U(LV),S(LV),F(3)
2070          F(1)=COS(U(1))
2080          F(2)=-SIN(U(1))
2090          F(3)=-COS(U(1))
2100          CALL DER(F,U,S)
2110          RETURN
```

```
2120          END
2130          SUBROUTINE SSIN(U,S)
2140          COMMON L,X1D,ID,L1,LV,IV
2150          DIMENSION U(LV),S(LV),F(3)
2160          F(1)=SIN(U(1))
2170          F(2)=COS(U(1))
2180          F(3)=-SIN(U(1))
2190          CALL DER(F,U,S)
2200          RETURN
2220          END
2260          SUBROUTINE AEXP(U,S)
2265          COMMON L,X1D,ID,L1,LV,IV
2270          DIMENSION U(LV),S(LV),F(2)
2280          F(1)=EXP(U(1))
2290          F(2)=F(1)
2310          CALL DER(F,U,S)
2320          RETURN
2330          END
2340          SUBROUTINE ADD(D,E,G)
2345          COMMON L,X1D,ID,L1,LV,IV
2350          DIMENSION D(LV),E(LV),G(LV)
2360          DO 50 I=1,LV
2370    50    G(I)=D(I)+E(I)
2380          RETURN
2390          END
2400          SUBROUTINE SR(U,S)
2410          DIMENSION U(20),S(20),F(4)
2420          F(1)=U(1)^.5
2430          F(2)=.5*U(1)^(-.5)
2440          F(3)=-(.25)*U(1)^(-1.5)
2450          F(4)=(3./8.)*U(1)^(-2.5)
2460          CALL DER(F,U,S)
2470          RETURN
2480          END
2490          SUBROUTINE DER(F,U,S)
2495          COMMON L,X1D,ID,L1,LV,IV
2500          DIMENSION U(LV),S(LV),F(2)
2510          S(1)=F(1)
2520          DO 55 I=1,IV
2530    55    S(I+1)=F(2)*U(I+1)
2930          RETURN
2940          END
2950          SUBROUTINE DIV(U,B,F1)
2960          DIMENSION U(8),B(8),R(8),F1(8),F(3)
2970          F(1)=U(1)^(-1.)
2980          F(2)=-U(1)^(-2.)
2990          F(3)=2.*U(1)^(-3.)
3010          CALL DER(F,U,R)
3015          CALL MULT(R,B,F1)
3020          RETURN
3030          END
3180          SUBROUTINE INTEG(X,HA,X1,U,U1,Z,DZ,IN,H,K,J,AH,AI,AJ,AK,Y)
3185          COMMON L,X1D,ID,L1,LV,IV
3190          DIMENSION U(L1),U1(L1),HA(LV),X(IV)
3195          DIMENSION Z(ID),DZ(ID),M(10)
3200          DIMENSION AH(ID),AI(ID),AJ(ID),AK(ID),Y(ID)
```

```
3205          DIMENSION X1(101,12)
3210          L1=L+1
3220          C1=1.0
3230          H2=H/2.0
3235          H1=ABS(H)
3260          I1=1
3265          XA=0.0
3267          X(1)=XA
3280    88    FORMAT(38H       X               Z1                 Z2)
3300          DO 90 I=1,K
3310          M(I)=I
3315    90    CONTINUE
3320          DO 98 N=1,L
3325          NL=N
3330          XS=XA+(N-1)*H
3340          X(1)=XS
3350          CALL MMULT(Z,C1,Y)
3355          DX=0.0
3360          CALL FORCE(X,HA,X1,U,U1,Z,DZ,IN,NL,DX)
3370    97    CALL MMULT(DZ,H2,AH)
3380          CALL MADD(Y,AH,Z)
3390          X(1)=XS+H2
3395          DX=0.5
3400          CALL FORCE(X,HA,X1,U,U1,Z,DZ,IN,NL,DX)
3410          CALL MMULT(DZ,H2,AI)
3420          CALL MADD(Y,AI,Z)
3430          CALL FORCE(X,HA,X1,U,U1,Z,DZ,IN,NL,DX)
3440          CALL MMULT(DZ,H,AJ)
3450          CALL MADD(Y,AJ,Z)
3460          X(1)=XS+H
3465          DX=1.0
3470          CALL FORCE(X,HA,X1,U,U1,Z,DZ,IN,NL,DX)
3475          CALL MMULT(DZ,H,AK)
3480          CALL MMUAD(AH,AI,AJ,AK,Y,Z)
3490          M1=M(I1)
3510          I1+I1+1
3515    92    IF(IN.EQ.2)GO TO 98
3520          DO 320 I=1,ID
3525   320    X1(N+1,I)=Z(I)
3530    98    CONTINUE
3535          RETURN
3540          END
3550          SUBROUTINE MMULT(Z,C1,Y)
3555          COMMON L,XID,ID,L1,LV,IV
3560          DIMENSION Z(ID),Y(ID)
3570          DO 94 I=1,ID
3580    94    Y(I)=C1*Z(I)
3590          RETURN
3600          END
3610          SUBROUTINE MADD(Y,AH,Z)
3615          COMMON L,XID,ID,L1,LV,IV
3620          DIMENSION Y(ID),AH(ID),Z(ID)
3630          DO 100 I=1,ID
3640   100    Z(I)=Y(I)+AH(I)
3650          RETURN
3660          END
```

```
3670          SUBROUTINE MMUAD(AH,AI,AJ,AK,Y,Z)
3675          COMMON L,X1D,ID,L1,LV,IV
3680          DIMENSION AH(ID),AI(ID),AJ(ID),AK(ID)
3685          DIMENSION Y(ID),Z(ID)
3690          DO 105 I=1,ID
3700    105   Z(I)=Y(I)+(2.*AH(I)+4.*AI(I)+2.*AJ(I)+AK(I))/6.
3710          RETURN
3720          END
3730          SUBROUTINE FORCE(X,HA,X1,U,U1,Z,DZ,IN,N,DX)
3740          COMMON L,X1D,ID,L1,LV,IV
3745          DIMENSION X1(L1,ID),U(L1),U1(L1)
3748          DIMENSION F1(18),F2(18),F3(18),F4(18)
3750          DIMENSION Z(ID),DZ(ID),X(IV)
3755          DIMENSION B(4,4)
3775          DO 300 I=1,ID
3780          I1=I+1
3785    300   X(I1)=Z(I)
3790          CALL INPUT(X,F1,F2,F3,F4)
3795          DZ(1)=F1(1)
3800          DZ(2)=F2(1)
3805          DZ(3)=F3(1)
3810          DZ(4)=F4(1)
3815          DO 310 I=1,4
3820          B(1,I)=F1(I+2)
3825          B(2,I)=F2(I+2)
3830          B(3,I)=F3(I+2)
3835          B(4,I)=F4(I+2)
3840          DZ(I+4)=0.0
3845    310   DZ(I+8)=0.0
3855          DO 330 J=1,4
3860          DO 340 I=1,4
3865          DZ(J+4)=DZ(J+4)+B(J,I)*Z(I+4)
3870    340   DZ(J+8)=DZ(J+8)+B(J,I)*Z(I+8)
3875    330   CONTINUE
3890    210   RETURN
3900          END
3910          SUBROUTINE TFORCE(TZ,DTZ,Z,IT)
3920          COMMON L,X1D,ID,L1,LV,IV
3930          DIMENSION TZ(ID),DTZ(ID),Z(IT)
3935          DIMENSION F1(18),F2(18),F3(18),F4(18),X(17)
3940          DO 320 I=1,ID
3945          I1=I+1
3950    320   X(I1)=Z(I)
3955          X(2)=TZ(1)
3960          X(3)=TZ(2)
3965          X(4)=0.0
3970          X(5)=0.0
3975          CALL INPUT(X,F1,F2,F3,F4)
3980          DEL=Z(7)*Z(12)-Z(8)*Z(11)
3985          DTZ(3)=(F4(1)*Z(11)-F3(1)*Z(12))/DEL
3990          DTZ(4)=(F3(1)*Z(8)-F4(1)*Z(7))/DEL
3995          DTZ(1)=F1(1)+Z(5)*DTZ(3)+Z(9)*DTZ(4)
4000          DTZ(2)=F2(1)+Z(6)*DTZ(3)+Z(10)*DTZ(4)
4005          RETURN
4015          END
```

Exercises

1. Derive Sukhanov's equations (5.28) and (5.29) for the second-order system two-point boundary value problem.

2. Derive Sukhanov's equations (5.56)–(5.59) for the fourth-order system two-point boundary value problem.

3. Given the fourth-order two-point boundary value problem

$$
\begin{aligned}
\dot{x}_1(t) &= x_2(t), & x_1(0) &= 1,\\
\dot{x}_2(t) &= p_2(t), & x_2(0) &= 0,\\
\dot{p}_1(t) &= x_1(t), & p_1(1) &= 0,\\
\dot{p}_2(t) &= -p_1(t), & p_2(1) &= 0,
\end{aligned}
$$

(a) Write the equations for Sukhanov's method for this problem.

(b) Obtain the numerical solution using the automatic derivative evaluation program.

Answer: See page 73, Reference 2.

6

Nonlinear Integral Equations

Integral equations appear in many engineering and physics problems. Numerical methods of solution for integral equations have been largely developed within the last 20 years (References 1–4). In this chapter a development involving an imbedding method for obtaining the numerical solution of nonlinear integral equations is described (References 5, 6). The numerical solution is obtained automatically from the initial value imbedding equations via the automatic derivative evaluation method described in the previous chapters (Reference 7). The derivatives required for the solution are computed automatically. The user need only enter the two known functions in equation (6.1) into the program. None of the derivatives associated with the imbedding method need be derived by hand. The subroutines are written in BASIC here, rather than in FORTRAN, because the former has several advantages in the manipulation of the vectors and matrices which occur in using the method of lines. The vectors consisting of the variables or functions of the variables, and all their derivatives are defined as derivative vectors (instead of just vectors) in this chapter to distinguish them from the other vectors and matrices.

The derivation of the imbedding equations is given in Section 6.1. This is followed by a discussion of the method of computation and the automatic derivative evaluation subroutines in Sections 6.2 and 6.3, respectively. Examples with numerical results are given in Section 6.4, and the BASIC program listing is given in Section 6.5.

6.1. Derivation of the Imbedding Equations

Consider the family of nonlinear integral equations (Reference 5)

$$F(u(t), t, \lambda) = \lambda \int_0^1 k(t, y, u(y))\, dy, \qquad 0 \le t \le 1, \qquad 0 \le \lambda \le \Lambda. \tag{6.1}$$

Equation (6.1) is a more general form of the linear integral equation

$$u(t) = g(t) + \lambda \int_0^1 k(t, y) u(y)\, dy \tag{6.2}$$

where $u(t)$ is an unknown function, $k(t, y)$ is the kernel, λ is the parameter of the integral equation, and $g(t)$ is the forcing function. For the linear case,

$$F(u(t), t, \lambda) = u(t) - g(t). \tag{6.3}$$

Since we intend to study the dependence of the solution u on the parameter λ, as well as on t, let $u(t) = u(t, \lambda)$ and express equation (6.1) in the form

$$F(u(t, \lambda), t, \lambda) = \lambda \int_0^1 k(t, y, u(y, \lambda))\, dy \tag{6.4}$$

Differentiating both sides of equation (6.4) with respect to λ yields

$$F_u(u(t, \lambda), t, \lambda) u_\lambda(t, \lambda) + F_\lambda(u(t, \lambda), t, \lambda) = \int_0^1 k(t, y, u(y, \lambda))\, dy + \lambda \int_0^1 k_u(t, y, u(y, \lambda)) u_\lambda(y, \lambda)\, dy \tag{6.5}$$

where the subscripts denote the partial derivatives with respect to u and λ. Equation (6.5) can be expressed in the form

$$u_\lambda(t, \lambda) = a(t, \lambda) + \lambda \int_0^1 \frac{k_u(t, y, u(y, \lambda))}{F_u((u, \lambda), t, \lambda)} u_\lambda(y, \lambda)\, dy \tag{6.6}$$

where

$$a(t, \lambda) = \left[-F_\lambda(u(t, \lambda), t, \lambda) + \int_0^1 k(t, y, u(y, \lambda))\, dy \right] \Big/ F_u(u(t, \lambda), t, \lambda) \tag{6.7}$$

Equation (6.6) is regarded as a Fredholm integral equation for the function

$u_\lambda(t, \lambda)$, the resolvent, K, of which satisfies the integral equation

$$K(t, y, \lambda) = \frac{k_u(t, y, u(y, \lambda))}{F_u(u(t, \lambda), t, \lambda)} + \lambda \int_0^1 \frac{k_u(t, y', u(y', \lambda))}{F_u(u(t, \lambda), t, \lambda)} K(y', y, \lambda)\, dy' \tag{6.8}$$

The solution of equation (6.6) can be expressed in terms of the resolvent, $K(t, y, \lambda)$, as follows:

$$u_\lambda(t, \lambda) = a(t, \lambda) + \lambda \int_0^1 K(t, y, \lambda) a(y, \lambda)\, dy \tag{6.9}$$

Equation (6.9) is a partial differential-integral equation for the function $u(t, \lambda)$. To obtain the resolvent, differentiate equation (6.8) with respect to λ to yield

$$K_\lambda(t, y, \lambda) = d(t, y, \lambda) + \lambda \int_0^1 \frac{k_u(t, y', u(y', \lambda))}{F_u(u(t, \lambda), t, \lambda)} K_\lambda(y', y, \lambda)\, dy', \tag{6.10}$$

where

$$\begin{aligned} d(t, y, \lambda) = {}& \frac{d}{d\lambda}\left[\frac{k_u(t, y, u(y, \lambda))}{F_u(u(t, \lambda), t, \lambda)}\right] + \int_0^1 \frac{k_u(t, y', u(y', \lambda))}{F_u(u(t, \lambda), t, \lambda)} K(y', y, \lambda)\, dy' \\ & + \lambda \int_0^1 \frac{d}{d\lambda}\left[\frac{k_u(t, y', u(y', \lambda))}{F_u(u(t, \lambda), t, \lambda)}\right] K(y', y, \lambda)\, dy' \end{aligned} \tag{6.11}$$

The derivative with respect to λ in the above equation is given by

$$\begin{aligned} \frac{d}{d\lambda}\left[\frac{k_u(t, y, u(y, \lambda))}{F_u(u(t, \lambda), t, \lambda)}\right] = {}& \{F_u(u(t, \lambda), t, \lambda) k_{uu}(t, y, u(y, \lambda)) u_\lambda(y, \lambda) \\ & - k_u(t, y, u(y, \lambda))[F_{uu}(u(t, \lambda), t, \lambda) u_\lambda(t, \lambda) \\ & + F_{u\lambda}(u(t, \lambda), t, \lambda)]\}[F_u(u(t, \lambda), t, \lambda)^2]^{-1} \end{aligned} \tag{6.12}$$

Equation (6.10) is itself a Fredholm integral equation for the function $K_\lambda(t, y, \lambda)$ and can be solved in terms of the resolvent,

$$K_\lambda(t, y, \lambda) = d(t, y, \lambda) + \lambda \int_0^1 K(t, y', \lambda) d(y', y, \lambda)\, dy'. \tag{6.13}$$

Equations (6.9) and (6.13), along with (6.7), (6.11), and (6.12), are the desired imbedding equations. The initial conditions at $\lambda = 0$ are

$$u(t, 0) = c(t), \tag{6.14}$$

$$K(t, y, 0) = \frac{k_u(t, y, c(y))}{F_u(c(t), t, 0)}, \qquad 0 \le t \le 1, \tag{6.15}$$

where the function $c(t)$ is determined from the condition

$$F(c(t), t, 0) = 0. \tag{6.16}$$

The solution of integral equation (6.1) is obtained by simultaneously integrating equations (6.9) and (6.13) with initial conditions (6.14) and (6.15).

6.2. Method of Computation

The numerical solution is obtained using the method of lines (Reference 8). The integrals are approximated in the interval (0, 1) by a quadrature formula of the form

$$\int_0^1 f(y)\, dy \cong \sum_{j=1}^{N} f(y_j) w_j, \tag{6.17}$$

where w_j are the weights from Simpson's rule. Introducing the nomenclature

$$K_{\lambda ij} = K_\lambda(t_i, y_j, \lambda), \tag{6.18}$$

$$u_{\lambda i} = u_\lambda(t_i, \lambda), \tag{6.19}$$

where the interval (0, 1) is divided into $(N - 1)$ equal parts of width $1/(N - 1)$ and

$$t_i = (i - 1)/(N - 1), \qquad i = 1, 2, \ldots, N, \tag{6.20}$$

$$y_j = (j - 1)/(N - 1), \qquad j = 1, 2, \ldots, N. \tag{6.21}$$

Equations (6.9) and (6.13) then become

$$u_{\lambda i} = a_i + \lambda \sum_{j=1}^{N} K_{ij} a_j w_j, \tag{6.22}$$

$$K_{\lambda ij} = d_{ij} + \lambda \sum_{m=1}^{N} K_{im} d_{mj} w_m. \tag{6.23}$$

In addition, define

$$K1_{ij} = \frac{k_u(t_i, y_j, u(y_j, \lambda))}{F_u(u(t_i, \lambda), t_i, \lambda)}, \tag{6.24}$$

$$K1_{\lambda ij} = \frac{d}{d\lambda} \left[\frac{k_u}{F_u} \right]_{ij}. \tag{6.25}$$

Equations (6.7), (6.11), and (6.12) then become

$$a_i = \left(-F_{\lambda i} + \sum_{j=1}^{N} k_{ij}w_j\right) \Big/ F_{ui}, \tag{6.26}$$

$$d_{ij} = K1_{\lambda ij} + \sum_{m=1}^{N} K1_{im}K_{mj}w_m + \lambda \sum_{m=1}^{N} K1_{\lambda im}K_{mj}w_m, \tag{6.27}$$

$$K1_{\lambda ij} = [F_{ui}k_{uuij}u_{\lambda j} - k_{uij}(F_{uui}u_{\lambda i} + F_{u\lambda i})]/F_{ui}^2. \tag{6.28}$$

The $N \times N$ matrices K_λ and d can be expressed in matrix form as the sum and product of matrices from equations (6.23) and (6.27) as follows:

$$K_\lambda = d + (\lambda)KWd, \tag{6.29}$$

$$d = K1_\lambda + K1WK + (\lambda)K1_\lambda WK, \tag{6.30}$$

where K, $K1$, and $K1_\lambda$ are $N \times N$ matrices, λ is a scalar parameter, and W is the $N \times N$ diagonal matrix

$$W = \text{diag}(w_1, w_2, \ldots, w_N). \tag{6.31}$$

6.3. Automatic Derivative Evaluation

In order to obtain the automatic solution of the nonlinear integral equation problem, the user of the program need only enter into the program the functions $F(u(t, \lambda), t, \lambda)$, and the integrand, $k(t, y, u(y, \lambda))$ defined in equation (6.4). The user must also specify the initial conditions and the upper integration limit Λ for the integration of u_λ and K_λ from $\lambda = 0$ to $\lambda = \Lambda$. For example, to evaluate the Ambarzumian integral equation, to be considered in Section 6.4, the user need only input the functional relationships

$$F(u(t, \lambda), t, \lambda) = 1 - u(t, \lambda)^{-1}, \tag{6.32}$$

$$k(t, y, u(y, \lambda)) = \frac{t}{2(t + y)} u(y, \lambda) \tag{6.33}$$

and specify the initial conditions and Λ. The program will then automatically evaluate the resolvent, $K(t, y, \lambda)$, and the unknown function, $u(t, \lambda)$.

Equations (6.9) and (6.13), along with (6.7), (6.11), and (6.12) show that the derivatives to be evaluated are F_u, F_λ, $F_{u\lambda}$, F_{uu}, k_u, and k_{uu}. Thus for each element of the vector $\mathbf{F}_i$, $i = 1, \ldots, N$, and the matrix k_{ij}, i, j

Table 6.1. Number of Components of the Derivative Vectors for Each of the Elements F_i or k_{ij}

Function	Number of components
$F(u(t, \lambda), t, \lambda)$ or $k(t, y, u(y, \lambda))$	1
First derivatives	4
Second derivatives	2[a]

[a] There are ten combinations of the second derivatives, but only two are required to evaluate the nonlinear integral equation problem.

$= 1, \ldots, N$, in equations (6.22)–(6.23) and (6.26)–(6.28), a derivative vector is evaluated containing the value of the element and the values of all of its derivatives. The dimension of the derivative vectors is obtained as follows. The first derivatives are with respect to t, y, u, and λ. The number of combinations of second derivatives is 10; however, only two are required, and are with respect to $u\lambda$ and uu. Thus the number of components of the derivative vectors is seven as summarized in Table 6.1. The definitions of the derivative vector components are given in Table 6.2.

In order to compute the derivatives automatically, the seven component derivative vectors must be computed for each of the four variables, t, y, u, and λ, and for each of the functions of the four variables, such as the sum, the product, the quotient, the square root, etc. The FORTRAN subroutines required to compute these vectors were described in Chapter 3.

Table 6.2. Definitions of the Derivative Vector Components

Vector component L	Derivative arguments kl	Symbol[a]
1	—	z
2	1	z_t
3	2	z_y
4	3	z_u
5	4	z_λ
6	34	$z_{u\lambda}$
7	33	z_{uu}

[a] $z_k = \partial z/\partial k$, $z_{kl} = \partial^2 z/\partial k \partial l$ where z is a scalar variable equal to t, y, u, λ or a function of the variables such as $F(u(t, \lambda), t, \lambda)$ or $k(t, y, u(y, \lambda))$.

In this chapter we will describe the subroutines using the BASIC language. With a few exceptions, the BASIC and FORTRAN subroutines are similar. The major differences will be emphasized in the following paragraphs.

6.3.1. Subroutine Linear

In subroutine LIN, the variables t, y, $u(y)$, and λ are represented by $S(I)$, $S(J)$, $R(J, N + 1)$, and X, respectively, and the corresponding derivative vectors are represented by **T**8, **Y**8, **U**8, and **L**8, as shown in Table 6.3. The first derivatives are equal to unity, while all the other derivatives are equal to zero. From Table 6.2

$$\mathbf{T}8(1) = t_i, \qquad \mathbf{Y}8(1) = y_j, \qquad \mathbf{U}8(1) = u(y_j), \qquad \mathbf{L}8(1) = \lambda, \tag{6.34}$$

$$\mathbf{T}8(2) = 1, \qquad \mathbf{Y}8(3) = 1, \qquad \mathbf{U}8(4) = 1, \qquad \mathbf{L}8(5) = 1, \tag{6.35}$$

and all other components of the seven component vectors, **T**8, **Y**8, **U**8, and **L**8 are equal to zero.

The vector **D**8 is used to represent a constant. The first component is equal to the constant and the rest of the components, corresponding to the derivatives, are equal to zero.

The order of the quadrature formula in equation (6.17) is N and is equal to 7 for the numerical results given in Section 6.4. The dimensions of the derivative vectors, described above, are coincidentally also equal to 7. To obtain greater numerical accuracy, the quadrature order can be increased by increasing N and redimensioning the appropriate vectors and matrices. The variables t and y are divided into $(N - 1)$ equal parts in the interval (0, 1) and are represented in the BASIC program by the variables

Table 6.3. BASIC Representation of the Variables

Variables	BASIC program	
	Variables	Vectors
t	$S(I)$	**T**8
y	$S(J)$	**Y**8
$u(y)$	$R(J, N + 1)$	**U**8
λ	X	**L**8

$S(I)$ and $S(J)$, i.e.,

$$S(I) = t_i = 0,\ 1/6,\ 1/3,\ 1/2,\ 2/3,\ 5/6,\ 1, \tag{6.36}$$

$$S(J) = y_i = 0,\ 1/6,\ 1/3,\ 1/2,\ 2/3,\ 5/6,\ 1, \tag{6.37}$$

$$I, J = 1, 2, \ldots, 7, \qquad i = I, \qquad j = J.$$

The resolvent $K(t_i, y_j, \lambda)$ is represented in the program by a partitioned 7×7 submatrix of the 7×8 matrix R, the $N+1 = 8$th column of which represents the unknown function $u(y_j)$ as shown in Table 6.3. To obtain $u(t_i)$, $R(J, N+1)$ is simply replaced by $R(I, N+1)$. The complete functions, $K(t_i, y_j, \lambda)$ and $u(t_i)$, for the seven values of t and y are obtained at each step interval, $\Delta\lambda$, by the integration of the partitioned matrix

$$F = [K_\lambda(t_i, y_j, \lambda) \mid u_\lambda(t_i, \lambda)], \tag{6.38}$$

where the 7×8 matrix F in the program represents the derivatives of the resolvent and the unknown function with respect to λ.

6.3.2. Programming the Derivative Vector Operations in BASIC

A subroutine in BASIC is entered with a GOSUB X statement, where X is the line number of the first statement in the subroutine. The last line in the subroutine is the RETURN statement. Since the line number, X, by itself does not give the name or a description of the subroutine to be called, a remarks statement, REM, can be used to give a name to the subroutine.

When a subroutine is called in FORTRAN, its dummy arguments are replaced by the corresponding actual arguments supplied in the CALL statement. BASIC subroutines do not have similar dummy arguments. Dummy arguments can be easily created if desired, however, by defining them each time before entering the GOSUB X statement. For example, for the derivative vector subroutines, ADD, MULT, and DIV, the seven component input vectors to be added, multiplied, or divided are always **A** and **B** and the output vector is always **C**. Operations involving vectors or matrices in BASIC are identified by the starting word, MAT.

An example of the call ADD statements and the subroutine ADD are given in Table 6.4. Lines 1380 and 1390 set the derivative vectors **T**8 and **Y**8 equal to the vectors **A** and **B**, respectively. Line 1410 calls the ADD subroutine. Lines 1940 to 1960 add the two vectors. For this example, the

Table 6.4. Example of the Call ADD Statements and the Subroutine ADD Using BASIC

```
1380    MAT A = T8
1390    MAT B = Y8
1400    Rem Call ADD
1410    Gosub 1940
 .
 .
 .
1930    Rem Subr ADD
1940    For I1 = 1 to N8
1950    Let C(I1) = A(I1) + B(I1)
1960    Next I1
1970    Return
```

first component of the vector **C** is

$$\begin{aligned} C(1) &= A(1) + B(1) \\ &= t_i + y_j. \end{aligned} \tag{6.39}$$

The other six components of the vector **C** correspond to the first and second derivatives given in Table 6.2.

In subroutine MULT the two functions **A** and **B** are multiplied:

$$C(1) = A(1)B(1) \tag{6.40}$$

The first derivatives are

$$C_k = A_k B + A B_k \tag{6.41}$$

where the subscripts denote the partial derivatives with respect to $k = 1$, 2, 3, and 4, corresponding to the variables, t, y, u, and λ. The second derivatives are

$$C_{u\lambda} = A_{u\lambda}B + A_u B_\lambda + A_\lambda B_u + AB_{u\lambda}, \tag{6.42}$$

$$C_{uu} = A_{uu}B + 2A_u B_u + AB_{uu}. \tag{6.43}$$

The program computes the first and second derivatives given by Equations (6.41)–(6.43) as follows:

$$C(I1 + 1) = A(I1 + 1)B(1) + A(1)B(I1 + 1), \qquad I1 = 1, \ldots, 4, \tag{6.44}$$

$$C(6) = A(6)B(1) + A(4)B(5) + A(5)B(4) + A(1)B(6), \tag{6.45}$$

$$C(7) = A(7)B(1) + 2A(4)B(4) + A(1)B(7). \tag{6.46}$$

In subroutine DIV, the derivatives of the quotient, B/A, are obtained. The derivatives of the reciprocal, $C2 = 1/A$, are evaluated first. Define a vector, **C**1, equal to the function $f(A)$ and its derivatives with respect to A as follows:

$$C1(1) = f(A) = A^{-1}, \tag{6.47}$$

$$C1(2) = f'(A) = -A^{-2}, \tag{6.48}$$

$$C1(3) = f''(A) = 2A^{-3}. \tag{6.49}$$

Then call subroutine DER to evaluate the function and its first and second derivatives with respect to t, y, u, and λ:

$$C2 = f(A), \tag{6.50}$$

$$C2_k = f'(A)A_k, \qquad k = 1, \ldots, 4, \tag{6.51}$$

$$C2_u = f''(A)A_\lambda A_u - f'(A)A_{u\lambda}, \tag{6.52}$$

$$C2_{uu} = f''(A)A_u^2 - f'(A)A_{uu}. \tag{6.53}$$

The program computes the function and its first and second derivatives given by equations (6.51)–(6.53) as follows:

$$C2(1) = C1(1), \tag{6.54}$$

$$C2(I1+1) = C1(2)A(I1+1), \qquad I1 = 1, \ldots, 4, \tag{6.55}$$

$$C2(6) = C1(3)A(4)A(5) + C1(2)A(6), \tag{6.56}$$

$$C2(7) = C1(3)A(4)^2 + C1(2)A(7). \tag{6.57}$$

Subroutine DIV then calls subroutine MULT to multiply $C2$ times B to obtain $C = C2*B = B/A$ and its derivatives.

6.4. Examples of Integral Equation Problems

Two examples of integral equation problems are given in this section. The first is a linear problem involving the buckling loads of columns and the second is a nonlinear problem involving the Ambarzumian integral equation. In both cases it is shown how to enter the problem into the automatic solution program via subroutine INPUT. Numerical results are given and compared with other methods of solution.

6.4.1. Linear Integral Equation

In References 4 and 9 it was shown that the differential equation of the deflection of a pin-ended column under axial load

$$\ddot{u}(t) + \lambda u(t) = f(t), \tag{6.58}$$

$$u(0) = 0, \qquad u(1) = 0 \tag{6.59}$$

can be represented by the equivalent integral equation

$$u(t) = g(t) + \lambda \int_0^1 k\big(t, y, u(y)\big)\, dy \tag{6.60}$$

where

$$k\big(t, y, u(y)\big) = \begin{cases} t(1 - y)u(y), & t \leq y \leq 1, \\ y(1 - t)u(y), & 0 \leq y \leq t. \end{cases} \tag{6.61}$$

The buckling or critical load of the column for $f(t) = g(t) = 0$ occurs at the singularities of the resolvent, i.e. at the smallest eigenvalue. This occurs at $\lambda = \pi^2$.

For this example we are not interested in the critical load of the column, but only in the value of the resolvent, $K(t, y, \lambda)$, at some convenient value of λ, i.e., at $\lambda = 2$, so that we can compare the numerical results with the exact analytical solution. In addition, in order to obtain values for $u(t)$, instead of setting $f(t) = 0$, we let

$$f(t) = 1 \tag{6.62}$$

in equation (6.58). This yields the value of $g(t)$ in equation (6.60),

$$g(t) = \tfrac{1}{2}t(t - 1). \tag{6.63}$$

Equation (6.60) is of the same form as equation (6.4). To obtain the automatic solution, the user enters the functions $F\big(u(t, \lambda), t, \lambda\big)$ and $k\big(t, y, u(y, \lambda)\big)$ into the program via subroutine INPUT. The function $k\big(t, y, u(y, \lambda)\big)$ is given by equation (6.61) and $F\big(u(t, \lambda), t, \lambda\big)$ is given by

$$F\big(u(t, \lambda), t, \lambda\big) = u(t, \lambda) - \tfrac{1}{2}t(t - 1). \tag{6.64}$$

Table 6.5 gives the BASIC listing of subroutine INPUT. The derivative vectors are evaluated for all the elements of the matrix k_{ij}, $i, j = 1, \ldots, N$ corresponding to the function $k\big(t, y, u(y, \lambda)\big)$ and all the components of the vector $\mathbf{F}_i$, $i = 1, \ldots, N$ corresponding to the function $F\big(u(t, \lambda), t, \lambda\big)$. The

Table 6.5. BASIC Listing of Subroutine INPUT for the Linear Integral Equation Example

Listing	Purpose
Rem Subr INPUT	
For I = 1 to N	
For J = 1 to N	
Gosub 1830	Call LIN. Defines the seven component derivative vectors corresponding to t, y, u, and λ.
If S(J) > S(I) then 1	If $y > t$ go to 1.
MAT A = D8	$D8(1) = 1.0$
MAT B = (−1)T8	
Gosub 1940	Call ADD. Forms the sum $C = 1 - t$.
MAT A = Y8	
MAT B = C	
Gosub 1990	Call MULT. Multiplies y by $(1 - t)$ to form $C = y(1 - t)$.
MAT A = U8	
MAT B = C	
Gosub 1990	Call MULT. Multiplies $u(y)$ times $y(1 - t)$ to form $C = y(1 - t)u(y)$.
Go to 2	
1 MAT A = D8	
MAT B = (−1)Y8	
Gosub 1940	Call ADD. Forms the sum $C = 1 - y$.
MAT A = T8	
MAT B = C	
Gosub 1990	Call MULT. Multiplies t times $(1 - y)$ to form $C = t(1 - y)$.
MAT A = U8	
MAT B = C	
Gosub 1990	Call MULT. Multiplies $u(y)$ times $t(1 - y)$ to form $C = t(1 - y)u(y)$.
2 Let E(I,J) = C(1)	Store $k(t, y, u(y))$.
Let E1(I,J) = C(4)	Store $k_u(t, y, u(y))$.
Let E2(I,J) = C(7)	Store $k_{uu}(t, y, u(y))$.
Next J	
Let U8(1) = R(I,N+1)	Set $u = u(t)$.
MAT F8 = U8	

Table 6.5. (*Continued*)

Listing	Purpose
Let P(I) = F8(1)	Store $F(t)$.
Let P1(I) = F8(4)	Store $F_u(t)$.
Let P2(I) = F8(5)	Store $F_\lambda(t)$.
Let P3(I) = F8(6)	Store $F_{u\lambda}(t)$.
Let P4(I) = F8(7)	Store $F_{uu}(t)$.
Next I	
For I = 1 to N	
For J = 1 to N	
Let K1(I,J) = E1(I,J)/P1(I)	Store k_u/F_u.
Next J	
Next I	
Return	

matrices k_{ij}, k_{uij}, and k_{uuij}, $i, j = 1, \ldots, N$, are stored following statement number 2 in Table 6.5. This is followed by setting the vector **F**8 equal to U8, corresponding to the functions $F(u(t, \lambda), t, \lambda)$ and $u(t, \lambda)$, respectively. The vectors $\mathbf{F}_i$, $\mathbf{F}_{ui}$, $\mathbf{F}_{\lambda i}$, $\mathbf{F}_{u\lambda i}$, and $\mathbf{F}_{uui}$, $i = 1, \ldots, N$ are then stored. While $F(u(t, \lambda), t, \lambda)$ is given by equation (6.64), the term $\frac{1}{2}t(t - 1)$ can be neglected in subroutine INPUT, since none of the derivatives with respect to t are required. Equations (6.22)–(6.23) and (6.26)–(6.28) are formed in subroutine FORCE from these stored values. Subroutine FORCE is used in the integration of equations (6.22)–(6.23).

The initial conditions are obtained by setting equation (6.64) equal to zero and solving for u. This yields

$$u(t, 0) = \tfrac{1}{2}t(t - 1). \tag{6.65}$$

This equation is entered into the program before subroutine INPUT is called for the first time. Subroutine INPUT then automatically yields the initial condition for the resolvent,

$$K(t, y, 0) = \frac{k_u(t, y, u(t, 0))}{F_u(u(t, 0), t, 0)}. \tag{6.66}$$

Table 6.6. Resolvent, $K(t, y, \lambda)$, of the Linear Integral Equation Example for $\lambda = 2$, $\Delta\lambda = 1/20$, $N = 7$

t	y						
	0	1/6	1/3	1/2	2/3	5/6	1
0	0	0	0	0	0	0	0
1/6	0	0.157536	0.136737	0.110873	0.076796	0.039875	0
1/3	0	0.136737	0.263345	0.213533	0.147903	0.076796	0
1/2	0	0.110873	0.213533	0.308284	0.213533	0.110873	0
2/3	0	0.076796	0.147903	0.213533	0.263345	0.136737	0
5/6	0	0.039875	0.076796	0.110873	0.136737	0.157536	0
1	0	0	0	0	0	0	0

The numerical results were obtained with the order of the quadrature N equal to 7 in equation (6.17). Differential equations (6.22) and (6.23) were integrated from $\lambda = 0$ to $\lambda = 2$ using a fourth-order Runge–Kutta method with grid intervals $\Delta\lambda = 1/20$. The numerical results for the resolvent, $K(t, y, \lambda)$, and for the solution of the unknown function, $u(t, \lambda)$, are given in Tables 6.6 and 6.7, respectively.

The exact analytical expression for the resolvent is given by

$$K(t, y, \lambda) = \begin{cases} (1/a)[\cos ay - \sin ay/\tan a] \sin at, & 0 \le t \le y, \\ (1/a)[\cos at - \sin at/\tan a] \sin ay, & 0 \le y \le t, \end{cases} \tag{6.67}$$

where

$$a = \sqrt{\lambda}. \tag{6.68}$$

Table 6.7. Solution, $u(t, \lambda)$, of the Linear Integral Equation Example for $\lambda = 2$, $\Delta\lambda = 1/20$, $N = 7$

t	$u(t, \lambda)$
0	0
1/6	−0.0869694
1/3	−0.139719
1/2	−0.159516
2/3	−0.139719
5/6	−0.0869694
1	0

The maximum error for the resolvent in Table 6.6 is less than 2.1% when compared with the exact solution.

The analytical solution of $u(t, \lambda)$ with $g(t)$ given by equation (6.63) is

$$u(t, \lambda) = (1/a^2)\left(1 + \frac{\cos a - 1}{\sin a} \sin at - \cos at\right). \tag{6.69}$$

The maximum error for $u(t, \lambda)$ in Table 6.7 is less than 1.2% when compared with the exact solution.

6.4.2. Nonlinear Integral Equation

The Ambarzumian integral equation, which occurs in the study of radiative transfer, is given by (References 2 and 10)

$$F(u(t, \lambda), t) = \lambda \int_0^1 k(t, y, u(y, \lambda))\, dy, \qquad 0 \leq t \leq 1, \quad 0 \leq \lambda \leq 1, \tag{6.70}$$

where

$$F(u(t, \lambda), t) = 1 - u(t, \lambda)^{-1}, \tag{6.71}$$

$$k(t, y, u(y, \lambda)) = \frac{t}{2(t + y)} u(y, \lambda). \tag{6.72}$$

To obtain the automatic solution, the user enters the functions $F(u(t, \lambda), t)$ and $k(t, y, u(y, \lambda))$ into the program via subroutine INPUT. The BASIC listing of subroutine INPUT is given in Table 6.8. As in the previous example, the derivative vectors are evaluated for all elements of the matrix k_{ij}, $i, j = 1, \ldots, N$ and for all components of the vector $\mathbf{F}_i$, $i = 1, \ldots, N$. The matrices k_{ij}, k_{uij}, and k_{uuij}, $i, j = 1, \ldots, N$, and the vectors $\mathbf{F}_i$, $\mathbf{F}_{ui}$, $\mathbf{F}_{\lambda i}$, $\mathbf{F}_{u\lambda i}$, and $\mathbf{F}_{uui}$, $i = 1, \ldots, N$ are stored in subroutine INPUT and are then used in subroutine FORCE to form the functions given by equations (6.22)–(6.23) and (6.26)–(6.28). Subroutine FORCE is used in the integration of equations (6.22)–(6.23).

The initial conditions are obtained by setting equation (6.71) equal to zero and solving for u. This yields

$$u(t, 0) = 1. \tag{6.73}$$

Equation (6.73) is entered into the program before subroutine INPUT is

Table 6.8. BASIC Listing of Subroutine INPUT for the Nonlinear Integral Equation Example

Listing	Purpose
Rem Subr INPUT	
For I = 1 to N	
For J = 1 to N	
Gosub 1830	Call LIN. Defines the seven-component derivative vectors corresponding to t, y, u, and λ.
MAT A = T8 MAT B = Y8 Gosub 1940	Call ADD. Forms the sum $C = t + y$.
MAT A = C MAT B = T8 Gosub 2070	Call DIV. Divides t by $(t + y)$ to form $C = t/(t + y)$.
MAT A = D8	D8(1) = 1/2.
MAT B = C Gosub 1990	Call MULT. Multiplies 1/2 by $t/(t + y)$ to form $C = t/2(t + y)$.
MAT A = U8 MAT B = C Gosub 1990	Call MULT. Multiplies $u(y)$ by $t/2(t + y)$ to form $C = [t/2(t + y)]u(y)$.
Let E(I,J) = C(1)	Store $k(t, y, u(y))$.
Let E1(I,J) = C(4)	Store $k_u(t, y, u(y))$.
Let E2(I,J) = C(7)	Store $k_{uu}(t, y, u(y))$.
Next J	
Let U8(1) = R(I,N+1)	Set $u = u(t)$.
MAT A = U8 Let D8(1) = −1.0 MAT B = D8 Gosub 2070	Call DIV. Divides −1 by $u(t)$ to form $C = -1/u(t)$.
Let D8(1) = 1.0 MAT A = D8 MAT B = C Gosub 1940	Call ADD. Forms the sum $C = 1 - 1/u(t)$.
MAT F8 = C	$F(t) = 1 - 1/u(t)$.

Table 6.8. (*Continued*)

Listing	Purpose
Let P(I) = F8(1)	Store F(t).
Let P1(I) = F8(4)	Store $F_u(t)$.
Let P2(I) = F8(5)	Store $F_\lambda(t)$.
Let P3(I) = F8(6)	Store $F_{u\lambda}(t)$.
Let P4(I) = F8(7)	Store $F_{uu}(t)$.
Next I	
For I = 1 to N	
For J = 1 to N	
Let K1(I,J) = E1(I,J)/P1(I)	Store k_u/F_u.
Next J	
Next I	
Return	

called for the first time. Subroutine INPUT then automatically yields the initial condition for the resolvent

$$K(t, y, 0) = \frac{k_u(t, y, u(t, 0))}{F_u(u(t, 0), t, 0)}. \tag{6.74}$$

The numerical results were obtained with the order of the quadrature N equal to 7 in equation (6.17). Differential equations (6.22) and (6.23) were integrated from $\lambda = 0$ to $\lambda = 0.5$ using a fourth-order Runge–Kutta method with grid intervals $\Delta\lambda = 1/20$. The numerical results for the resolvent, $K(t, y, \lambda)$, and for the solution of the unknown function, $u(t, \lambda)$, are given in Tables 6.9 and 6.10, respectively.

For comparison, the solution in Reference 10 at $t = 1$, $\lambda = 0.5$ is

$$u(1, 0.5) = 1.251, \tag{6.75}$$

which agrees with the numerical results in Table 6.10 exactly. Also shown in Table 6.10 are the results for the integration from $\lambda = 0$ to $\lambda = 0.95$ with $\Delta\lambda = 1/20$ and $1/40$. It is seen that decreasing grid interval size from $\Delta\lambda = 1/20$ to $1/40$ changes the solution by less than 0.015%. The solution

Table 6.9. Resolvent, $K(t, y, \lambda)$, of the Nonlinear Integral Equation Example for $\lambda = 0.5$, $\Delta\lambda = 1/20$, $N = 7$

t	y						
	0	1/6	1/3	1/2	2/3	5/6	1
0	0	0	0	0	0	0	0
1/6	0.671554	0.346778	0.235911	0.179437	0.145047	0.121844	0.105104
1/3	0.771897	0.516273	0.389556	0.313273	0.262173	0.225501	0.19788
1/2	0.845341	0.625503	0.499011	0.415633	0.356324	0.311913	0.27739
2/3	0.902352	0.704341	0.581742	0.496365	0.433132	0.384311	0.345442
5/6	0.948204	0.764865	0.646843	0.561693	0.496811	0.445565	0.404001
1	0.986009	0.813192	0.699589	0.615668	0.550394	0.497935	0.45476

in Reference 10 at $t = 1$, $\lambda = 0.95$ is

$$u(1, 0.95) = 2.077, \tag{6.76}$$

which agrees with the numerical results in Table 6.10 to within an error of less than 0.25%. The accuracy can be increased by increasing the order of the quadrature.

Table 6.10. Solution, $u(t, \lambda)$, of the Nonlinear Integral Equation Example for $N = 7$

t	$u(t, \lambda)$		
	$\lambda = 0.5$ $\Delta\lambda = 1/20$	$\lambda = 0.95$ $\Delta\lambda = 1/20$	$\lambda = 0.95$ $\Delta\lambda = 1/40$
0	1	1	1
1/6	1.1017	1.29256	1.29258
1/3	1.15243	1.4967	1.49676
1/2	1.18758	1.66901	1.66912
2/3	1.21386	1.81925	1.81942
5/6	1.23444	1.95248	1.9527
1	1.25105[a]	2.07191	2.0722

[a] Reference 10 values are $u(1, 0.5) = 1.251$ and $u(1, 0.95) = 2.077$.

Table 6.11. Definitions of the BASIC Variables, Vectors, and Matrices for the Nonlinear Integral Equation Example[a]

	Variable	BASIC variable, vector, or matrix	Number of elements	Program line number
Initial conditions	$u(t, 0)$	$R(I, N+1)$	7×1	820
	$K(t, y, 0)$	**K1**	7×7	880
Order of the quadrature	N	N	1	170
Integration grid interval size	$\Delta\lambda$	H	1	160
Upper integration limit	Λ	L	1	150, 1320
Resolvent	$K(t, y, \lambda)$	**R1**	7×7	980
Unknown function	$u(t, \lambda)$	$R(I, N+1)$	7×1	1580, 1870
Partitioned matrix	$[K(t, y, \lambda) \vdots u(t, \lambda)]$	**R**	7×8	660
Derivatives to be integrated each iteration	$K_\lambda(t, y, \lambda)$	**F1**	7×7	1250
	$u_\lambda(t, \lambda)$	**U1**	7×1	1060
Partitioned matrix	$[K_\lambda(t, y, \lambda) \vdots u_\lambda(t, \lambda)]$	**F**	7×8	1070, 1280
Function $F(u(t, \lambda), t, \lambda)$ and its derivatives	$F(t)$	**P**	7×1	1700
	$F_u(t)$	**P1**	7×1	1710
	$F_\lambda(t)$	**P2**	7×1	1720
	$F_{u\lambda}(t)$	**P3**	7×1	1730
	$F_{uu}(t)$	**P4**	7×1	1740
Integrand $k(t, y, u(y, \lambda))$ and its derivatives	$k(t, y, u(y))$	**E**	7×7	1540
	$k_u(t, y, u(y))$	**E1**	7×7	1550
	$k_{uu}(t, y, u(y))$	**E2**	7×7	1560
Other functions	$a(t, \lambda)$	**L**	7×1	990
	$d(t, y, \lambda)$	**D**	7×7	1210
	k_u/F_u	**K1**	7×7	1780
	$\dfrac{d}{d\lambda}\left[\dfrac{k_u}{F_u}\right]$	**K2**	7×7	1130
Quadrature weights (Simpson's rule)	$\mathrm{diag}(w_1, \ldots, w_7)$	**W**	7×7	210–280

[a] Additional definitions of the variables and vectors are given in Tables 6.2 and 6.3.

6.5. Program Listing

The BASIC program listing for the Ambarzumian nonlinear integral equation example is given in this section. The definitions of the BASIC variables, vectors, and matrices are given in Table 6.11. Simpson's rule is used for the quadrature formula of order $N = 7$. The differential equations for the resolvent, $K(t, y, \lambda)$, and the function, $u(t, \lambda)$, are integrated simultaneously using a fourth-order Runge–Kutta method. A major advantage of using BASIC is the ease with which matrix operations, such as those used in the integration routine, can be performed. The integration routine is given in line numbers 430–680 of the program listing. The complete function, $u(t, \lambda)$ for the seven values of $t = 0, 1/6, 1/3, 1/2, 2/3, 5/6, 1$ is obtained at each step interval, $\Delta\lambda$, by the integration of $u_\lambda(t, \lambda)$.

The resolvent, $K(t, y, \lambda)$, is also available simultaneously for the 49 values of $t, y = 0, 1/6, 1/3, 1/2, 2/3, 5/6, 1$ at each step interval, $\Delta\lambda$, by the integration of $K_\lambda(t, y, \lambda)$, although not printed out. The quadrature order can be increased in order to obtain greater numerical accuracy, if desired, by increasing N and redimensioning the appropriate vectors and matrices.

```
10 DIM C1(3),C2(8),P2(7),P3(7),P4(7),K2(7,7),E2(7,7),V1(7,7)
20 DIM A(8),B(8),C(8)
30 DIM E(7,7),E1(7,7),F8(8),P(7),P1(7)
40 DIM O(7),O1(7)
50 DIM U(7),U1(7),L(7),R1(7,7),F1(7,7)
60 DIM S(7),Q(7,7),R(7,8),V(7,7),D(7,7)
70 DIM F(7,8),W(7,7),K1(7,7)
80 DIM Y(7,8),H(7,8),I(7,8),J(7,8),K(7,8)
90 LET N8=7
100 MAT A=ZER(N8)
110 MAT B=ZER(N8)
120 MAT C=ZER(N8)
130 MAT C1=ZER(3)
140 MAT C2=ZER(N8)
150 READ L
160 LET H=1/20
170 LET N=7
180 FOR I=0 TO N-1
190 LET S(I+1)=I/(N-1)
200 NEXT I
210 MAT W=ZER(N,N)
220 FOR I=1 TO ((N-1)/2-1)
230 LET W(2*I,2*I)=2/9
240 LET W(1+2*I,1+2*I)=1/9
250 NEXT I
260 LET W(1,1)=1/18
270 LET W(N,N)=1/18
280 LET W(N-1,N-1)=2/9
290 MAT R=ZER(7,8)
300 MAT T8=ZER(N8)
310 MAT Y8=ZER(N8)
```

```
320 MAT U8=ZER(N8)
330 MAT L8=ZER(N8)
340 MAT D8=ZER(N8)
350 LET X=0
360 GOSUB 810
370 PRINT "KERNEL="
380 PRINT "R,E,E1,K1,F8,P,P1="
390 MAT PRINT R;E;E1;K1;F8;P;P1;
400 PRINT "K2,E2,P2,P3,P4="
410 MAT PRINT K2;E2;P2;P3;P4;
420 PRINT
430 FOR N1=1 TO L/H
440 MAT Y=(1)*R
450 LET X=(N1-1)*H
460 GOSUB 940
470 MAT H=(H/2)*F
480 MAT R=Y+H
490 LET X=X+H/2
500 GOSUB 940
510 MAT I=(H/2)*F
520 MAT R=Y+I
530 GOSUB 940
540 MAT J=(H)*F
550 MAT R=Y+J
560 LET X=X+H/2
570 GOSUB 940
580 MAT K=(H)*F
590 MAT H=(2)*H
600 MAT I=(4)*I
610 MAT J=(2)*J
620 MAT I=H+I
630 MAT J=I+J
640 MAT K=J+K
650 MAT K=(1/6)*K
660 MAT R=Y+K
670 LET X=N1*H
680 NEXT N1
690 REM CALL INPUT
700 GOSUB 1340
710 PRINT "N1="N1,"LAMBDA="X
720 PRINT
730 PRINT "E,E1,K1,F8,P,P1"
740 MAT PRINT E;E1;K1;F8;P;P1
750 PRINT "K2,E2,P2,P3,P4="
760 MAT PRINT K2;E2;P2;P3;P4;
770 PRINT "R="
780 MAT PRINT R;
790 GO TO 2270
800 REM INITIAL COND
810 FOR I=1 TO N
820 LET R(I,N+1)=1
830 NEXT I
840 REM CALL INPUT
850 GOSUB 1340
860 FOR I=1 TO N
870 FOR J=1 TO N
```

```
880 LET R(I,J)=K1(I,J)
890 NEXT J
900 NEXT I
910 RETURN
920 RE SUBR FORCE
930 REM CALL INPUT
940 GOSUB 1340
950 FOR I=1 TO N
960 LET L(I)=-P2(I)
970 FOR J=1 TO N
980 LET R1(I,J)=R(I,J)
990 LET L(I)=L(I)+E(I,J)*W(J,J)
1000 NEXT J
1010 LET L(I)=L(I)/P1(I)
1020 NEXT I
1030 FOR I=1 TO N
1040 LET U1(I)=L(I)
1050 FOR J=1 TO N
1060 LET U1(I)=U1(I)+X*R1(I,J)*L(J)*W(J,J)
1070 LET F(I,N+1)=U1(I)
1080 NEXT J
1090 NEXT I
1100 FOR I=1 TO N
1110 FOR J=1 TO N
1120 LET K2(I,J)=P1(I)*E2(I,J)*U1(J)-E1(I,J)*(P4(I)*U1(I)+P3(I))
1130 LET K2(I,J)=K2(I,J)/(P1(I))^2
1140 NEXT J
1150 NEXT I
1160 MAT Q=W*R1
1170 MAT D=K1*Q
1180 MAT V=K2*Q
1190 MAT V1=(X)*V
1200 MAT D=D+V1
1210 MAT D=K2+D
1220 MAT V=W*D
1230 MAT Q=R1*V
1240 MAT V=(X)*Q
1250 MAT F1=D+V
1260 FOR I=1 TO N
1270 FOR J=1 TO N
1280 LET F(I,J)=F1(I,J)
1290 NEXT J
1300 NEXT I
1310 RETURN
1320 DATA 0.5
1330 REM SUBR INPUT
1340 FOR I=1 TO N
1350 FOR J=1 TO N
1360 REM CALL LIN
1370 GOSUB 1830
1380 MAT A=T8
1390 MAT B=YB
1400 REM CALL ADD
1410 GOSUB 1940
1420 MAT A=C
1430 MAT B=T8
```

```
1440 REM CALL DIV,C=B/A
1450 GOSUB 2070
1460 MAT A=D8
1470 MAT B=C
1480 REM CALL MULT
1490 GOSUB 1990
1500 MAT A=U8
1510 MAT B=C
1520 REM CALL MULT
1530 GOSUB 1990
1540 LET E(I,J)=C(1)
1550 LET E1(I,J)=C(4)
1560 LET E2(I,J)=C(7)
1570 NEXT J
1580 LET U8(1)=R(I,N+1)
1590 MAT A=U8
1600 LET D8(1)=-1.0
1610 MAT B=D8
1620 REM CALL DIV,C=B/A
1630 GOSUB 2070
1640 LET D8(1)=1.0
1650 MAT A=D8
1660 MAT B=C
1670 REM CALL ADD
1680 GOSUB 1940
1690 MAT F8=C
1700 LET P(I)=F8(1)
1710 LET P1(I)=F8(4)
1720 LET P2(I)=F8(5)
1730 LET P3(I)=F8(6)
1740 LET P4(I)=F8(7)
1750 NEXT I
1760 FOR I=1 TO N
1770 FOR J=1 TO N
1780 LET K1(I,J)=E1(I,J)/P1(I)
1790 NEXT J
1800 NEXT I
1810 RETURN
1820 REM SUBR LIN
1830 LET T8(1)=S(I)
1840 LET T8(2)=1.0
1850 LET Y8(1)=S(J)
1860 LET Y8(3)=1.0
1870 LET U8(1)=R(J,N+1)
1880 LET U8(4)=1.0
1890 LET L8(1)=X
1900 LET L8(5)=1.0
1910 LET D8(1)=0.5
1920 RETURN
1930 REM SUBR ADD
1940 FOR I1=1 TO N8
1950 LET C(I1)=A(I1)+B(I1)
1960 NEXT I1
1970 RETURN
1980 REM SUBR MULT
1990 LET C(1)=A(1)*B(1)
```

```
2000 FOR I1=1 TO 4
2010 LET C(I1+1)=A(I1+1)*B(1)+A(1)*B(I1+1)
2020 NEXT I1
2030 LET C(6)=A(6)*B(1)+A(4)*B(5)+A(5)*B(4)+A(1)*B(6)
2040 LET C(7)=A(7)*B(1)+2*A(4)*B(4)+A(1)*B(7)
2050 RETURN
2060 REM SUB DIV,C=B/A
2070 IF ABS(A(1))<1.E-6 THEN 2130
2080 LET C1(1)=A(1)^(-1)
2090 LET C1(2)=-A(1)^(-2)
2100 LET C1(3)=2*A(1)^(-3)
2110 REM CALL DER,C2=1/A
2120 GO TO 2140
2130 MAT C1=ZER
2140 GOSUB 2200
2150 MAT A=C2
2160 REM CALL MULT C=C2*B
2170 GOSUB 1990
2180 RETURN
2190 REM SUBR DER,C2=1/A
2200 C2(1)=C1(1)
2210 FOR I1=1 TO 4
2220 LET C2(I1+1)=C1(2)*A(I1+1)
2230 NEXT I1
2240 LET C2(6)=C1(3)*A(4)*A(5)+C1(2)*A(6)
2250 LET C2(7)=C1(3)*A(4)^2+C1(2)*A(7)
2260 RETURN
2270 END
```

Exercises

1. Verify that equation (6.67) is the analytical expression of the resolvent for the buckling load problem.

2. Verify that equation (6.69) is the analytical solution for the unknown function $u(t)$ when $f(t)$ is given by equation (6.62).

3. Program the automatic derivative evaluation method and find the resolvent and the unknown function $u(t)$ for the buckling load problem when $\lambda = 4$. Compare the results with the analytical solution.

References

Preface

1. WENGERT, R. E., A simple automatic derivative evaluation program, *Communications of the ACM*, Vol. 7, No. 8, pp. 463–464, August 1964.
2. BELLMAN, R., KAGIWADA, H., and KALABA, R., Wengert's numerical method for partial derivatives, orbit determination and quasilinearization, *Communications of the ACM*, Vol. 8, No. 4, pp. 231–232, April 1965. (This paper first appeared as research memorandum RM-4354, November 1964, the RAND Corporation, Santa Monica, California.)
3. BELLMAN, R., and KALABA, R., *Quasilinearization and Nonlinear Boundary Value Problems*, Elsevier, New York, 1965.

Chapter 1

1. WENGERT, R. E., A simple automatic derivative evaluation program, *Communications of the ACM*, Vol. 7, No. 8, pp. 463–464, August 1964.
2. WILKINS, R. D., Investigation of a new analytical method for numerical derivative evaluation, *Communications of the ACM*, Vol. 7, No. 8, pp. 465-471, August 1964.
3. BELLMAN R., and KALABA, R., *Quasilinearization and Nonlinear Boundary Value Patterns*, Elsevier, New York, 1965.
4. BELLMAN, R., KAGIWADA, H., and KALABA, R., Wengert's numerical method for partial derivatives, orbit determination and quasilinearization, *Communications of the ACM*, Vol. 8, No. 4, pp. 231–232, April 1965. (This paper first appeared as research memorandum RM-4354, November 1964, The RAND Corporation, Santa Monica, California.)
5. RALL, L. B., *Automatic Differentiation: Techniques and Applications*, Springer-Verlag, Berlin, 1981.

6. KALABA, R., TESFATSION, L., and WANG, J. L., A finite algorithm for the exact evaluation of higher-order partial derivatives of functions of many variables, *Journal of Mathematical Analysis and Applications*, Vol. 92, pp. 552–563, 1983.
7. WEXLER, A. S., Automatic evaluation of derivatives, submitted for publication.

Chapter 2

1. KALABA, R., RASAKHOO, N., and TISHLER, A., Nonlinear least squares via automatic derivative evaluation, *Applied Mathematics and Computation*, Vol. 12, pp. 119–137, 1983.
2. KALABA, R., and TISHLER, A., Automatic derivative evaluation in the optimization of nonlinear models, *The Review of Economics and Statistics*, Vol. 66, pp. 653–660, 1984.
3. KALABA, R., and TISHLER, A., On the use of the table method in constrained optimization, *Journal of Optimization Theory and Applications*, Vol. 43, No. 2, pp. 157–165, 1984.

Chapter 3

1. KALABA, R., and SPINGARN, K., *Control, Identification, and Input Optimization*, Plenum Press, New York, 1982.
2. ATHANS, M., and FALB, P., *Optimal Control*, McGraw-Hill, New York, 1966.
3. BRYSON, A. E., Jr., and HO, Y., *Applied Optimal Control*, Blaisdell, Waltham, Massachusetts, 1969.
4. SAGE, A. P., and WHITE, C. C., III., *Optimum Systems Control*, Prentice-Hall, Englewood Cliffs, New Jersey, 1977.
5. GOTTFRIED, B. S., and WEISSMAN, J., *Introduction to Optimization Theory*, Prentice-Hall, Englewood Cliffs, New Jersey, 1973.
6. PONTRYAGIN, L. S., *et al.*, *The Mathematical Theory of Optimal Processes*, Wiley-Interscience, New York, 1962.
7. GELFAND, I. M., and FOMIN, S. V., *Calculus of Variations*, Prentice-Hall, Englewood Cliffs, New Jersey, 1963.
8. COURANT, R., and HILBERT, D., *Methods of Mathematical Physics*, Vol. 1, Wiley-Interscience, New York, 1953.
9. MIELE, A., Introduction to the calculus of variations in one independent variable, in *Theory of Optimum Aerodynamic Shapes*, edited by A. Miele, Academic, New York, pp. 3–19, 1965.
10. LEONDES, C. T., editor of the series in *Control and Dynamic Systems*, *Advances in Theory and Applications*, Academic, New York, Volumes 1–21, 1965–1984.
11. WENGERT, R., A simple automatic derivative evaluation program, *Communications of the ACM*, Vol. 7, pp. 463–464, 1964.
12. BELLMAN, R., KAGIWADA, H., and KALABA, R., Wengert's numerical method for partial derivatives, orbit determination, and quasilinearization, *Communications of the ACM*, Vol. 8, pp. 231–232, 1965.

13. Kalaba, R., and Spingarn, K., Automatic solution of optimal control problems, I. Simplest problem in the calculus of variations, *Applied Mathematics and Computation*, Vol. 13, pp. 131–148, February 1984.
14. Kalaba, R., and Spingarn, K., Automatic solution of optimal control problems, II. First-order nonlinear systems, American Control Conference, San Francisco, California, June 1983.
15. Kalaba, R., and Spingarn, K., Automatic solution of optimal control problems, III. Differential and integral constraints, *IEEE Control Systems Magazine*, Vol. 4, pp. 3–8, February 1984.
16. Kalaba, R., and Spingarn, K., Automatic solution of optimal control problems, IV. Gradient method, *Applied Mathematics and Computation*, Vol. 14, pp. 289–300, April 1984.
17. Kalaba, R., and Spingarn, K., Automatic solution of optimal control problems, V. Second-order nonlinear systems, Seventeenth Asilomar Conference on Circuits, Systems, and Computers, Pacific Grove, California, November 1983.
18. Kalaba, R., and Spingarn, K., Automatic solution of Nth-order optimal control problems, *IEEE Transactions on Aerospace and Electronic Systems*, Vol. 21, pp. 345–350, May 1985.
19. Bellman, R. E., and Kalaba, R. E., *Quasilinearization and Nonlinear Boundary-Value Problems*, Elsevier, New York, 1965.
20. Mullins, E. R., Jr., and Rosen, D., *Probability and Calculus*, Bogden & Quigley, Tarrytown-on-the-Hudson, New York, p. 66, 1971.

Chapter 4

1. Bellman, R., and Kalaba, R., *Quasilinearization and Nonlinear Boundary Value Problems*, Elsevier, New York, 1965.
2. Kalaba, R., and Spingarn, K., *Control, Identification and Input Optimization*, Plenum, New York, 1982.
3. Kagiwada, H., *System Identification: Methods and Applications*, Addison-Wesley, Reading, Massachusetts, 1974.
4. Kalaba, R., Mazer, N., Murthy, V. K., Rasakhoo, N., and Spingarn, K., Semi-Markov models for target search in an image, *Proceedings of the 32nd National IRIS*, May 1984.
5. Kalaba, R., and Tishler, A., A generalized Newton algorithm to minimize a function with many variables using computer evaluated exact higher order derivatives, *Journal of Optimization Theory and Applications*, Vol. 42, No. 3, pp. 383–395, 1984.
6. Kalaba, R., and Tishler, A., On the use of the table method in constrained optimization, *Journal of Optimization Theory and Applications*, Vol. 43, No. 4, pp. 543–555, 1984.
7. Kalaba, R., Rasakhoo, N., and Tishler, A., Nonlinear least squares via automatic derivative evaluation, *Applied Mathematics and Computation*, Vol. 12, pp. 119–137, 1983.
8. Kalaba, R., Tesfatsion, L., and Wang, J. L., A finite algorithm for the exact evaluation of higher-order partial derivatives of functions of many variables, *Journal of Mathematical Analysis and Applications*, Vol. 92, pp. 552–563, 1983.

9. Kalaba, R., On nonlinear differential equations, the maximum operation and monotone convergence, *Journal of Mathematics and Mechanics*, Vol. 8, pp. 519–574, 1959.
10. Kalaba, R., Langetieg, T., Rasakhoo, N., and Weinstein, M., Estimation of implicit bankruptcy costs, *The Journal of Finance*, Vol. 39, pp. 629–642, 1984.
11. Rasakhoo, N., Fitting Contingent Claim Pricing Models to Data, Ph.D. Thesis, Department of Finance, University of Southern California, 1985.

Chapter 5

1. Sukhanov, A. A., A method of solution of nonlinear two-point boundary value problems, *Journal of Numerical Mathematics and Mathematical Physics*, Vol. 23, pages 228–231, 1983 (in Russian).
2. Kalaba, R., and Spingarn, K., *Control, Identification, and Input Optimization*, Plenum Press, New York, 1982.
3. Bellman, R. E., and Kalaba, R., *Quasilinearization and Nonlinear Boundary-Value Problems*, Elsevier, New York, 1965.
4. Kagiwada, H., Kalaba, R., Rasakhoo, N., and Spingarn, K., Numerical experiments using Sukhanov's initial-value method for nonlinear two-point boundary value problems, *Computers and Mathematics with Applications*, Vol. 10, Nos. 4/5, pp. 327–330, 1984.
5. Kagiwada, H., Kalaba, R., Rasakhoo, N., and Spingarn, K., Numerical experiments using Sukhanov's initial value method for nonlinear two-point boundary value problems II, *Journal of Optimization Theory and Applications*, Vol. 46, No. 1, May 1985.
6. Kagiwada, H., Kalaba, R., Rasakhoo, N., and Spingarn, K., Numerical experiments using Sukhanov's initial-value method for nonlinear two-point boundary value problems III, *Nonlinear Analysis, Theory, Methods, and Applications*, Vol. 8, No. 12, pp. 1497–1505, December 1984.
7. Kagiwada, H., Kalaba, R., Rasakhoo, N., and Spingarn, K., Numerical experiments using Sukhanov's initial-value method for nonlinear two-point boundary value problems IV, *Nonlinear Analysis, Theory, Methods, and Applications*, Vol. 9, No. 5, pp. 469–477, 1985.
8. Kagiwada, H., and Kalaba, R., Derivation and validation of an initial-value method for certain nonlinear two-point boundary value problems, *Journal of Optimization Theory and Applications*, Vol. 2, No. 6, pp. 378–385, 1968.

Chapter 6

1. Golberg, M. A., *Solution Methods for Integral Equations, Theory and Applications*, Plenum Press, New York, 1979.
2. Kagiwada, H. H., and Kalaba, R., *Integral Equations Via Imbedding Methods*, Addison-Wesley, Reading, Massachusetts, 1974.
3. Casti, J., and Kalaba, R., *Imbedding Methods in Applied Mathematics*, Addison-Wesley, Reading, Massachusetts, 1973.

4. KALABA, R., and SPINGARN, K., *Control, Identification, and Input Optimization*, Plenum Press, New York, 1982.
5. KAGIWADA, H. H., and KALABA, R., An initial value method for the nonlinear integral equation $F(u(t), t, \lambda) = \lambda \int_0^1 k(t, y, u(y))\,dy$, *Applied Mathematics and Computation*, Vol. 13, Nos. 1 and 2, pp. 117–124, August 1983.
6. KAGIWADA, H. H., KALABA, R., and SPINGARN, K., Automatic solution of nonlinear integral equations, *Computers and Mathematics with Applications* (in press).
7. KALABA, R., and SPINGARN, K., Automatic solution of *N*th-order optimal control problems, *IEEE Transactions on Aerospace and Electronic Systems*, Vol. 21, pp. 345–350, May 1985.
8. MIKHLIN, S., and SMOLITSKIY, K., *Approximate Methods for Solution of Differential and Integral Equations*, American Elsevier, New York, 1967.
9. KALABA, R. E., SPINGARN, K., and ZAGUSTIN, E., On the integral equation method for buckling loads, *Applied Mathematics and Computation*, Vol. 1, No. 3, pp. 253–261, 1975.
10. SOBOLEV, V. V., *Scattering of Light in Planetary Atmospheres*, Nauka, Moscow, 1972.

Additional References

1. VOLIN, YU. M., and OSTROVSKII, G. M., Automatic computation of derivatives with the use of multilevel differentiating technique. I. Algorithmic basis, *Computers and Mathematics with Applications*, Vol. 11, No. 11, pp. 1099-1114, 1985.
2. WILCOX, R., and HARTEN, L., Macsyma-generated closed form solutions to some matrix Riccati equations, *Applied Mathematics and Computation*, Vol. 14, pp. 149-166, 1984.
3. RICHERT, A., A non-Simpsonian use of parabolas in numerical integration, *American Mathematical Monthly*, Vol. 92, No. 6, pp. 425-426, 1985.

Author Index

Subject Index

ENCYCLOPEDIA BROWN

and the Case of the Jumping Frogs

Don't miss any of the other books about

ENCYCLOPEDIA BROWN!

Encyclopedia Brown, Boy Detective

Encyclopedia Brown and the Case of the Secret Pitch

Encyclopedia Brown Finds the Clues

Encyclopedia Brown Gets His Man

Encyclopedia Brown Solves Them All

Encyclopedia Brown Keeps the Peace

Encyclopedia Brown Saves the Day

Encyclopedia Brown Tracks Them Down

Encyclopedia Brown Shows the Way

Encyclopedia Brown Takes the Case

Encyclopedia Brown Lends a Hand

Encyclopedia Brown and the Case of the Dead Eagles

Encyclopedia Brown and the Case of the Midnight Visitor

Encyclopedia Brown and the
Case of the Mysterious Handprints

Encyclopedia Brown and the
Case of the Treasure Hunt

Encyclopedia Brown and the
Case of the Disgusting Sneakers

Encyclopedia Brown and the
Case of the Two Spies

Encyclopedia Brown and the Case of Pablo's Nose

Encyclopedia Brown and the
Case of the Sleeping Dog

Encyclopedia Brown and the
Case of the Slippery Salamander

Encyclopedia Brown and the
Case of the Jumping Frogs

ENCYCLOPEDIA BROWN

and the Case of the Jumping Frogs

DONALD J. SOBOL

Illustrated by Robert Papp

DELACORTE PRESS

Published by
Delacorte Press
an imprint of
Random House Children's Books
a division of Random House, Inc.
New York

Visit us on the Web! www.randomhouse.com/kids
Educators and librarians, for a variety of teaching tools, visit us at www.randomhouse.com/teachers

Cataloging-in-Publication data is available from the Library of Congress.
ISBN: 0-385-729316 (trade) 0-385-90148-8 (lib. bdg.)

The text of this book is set in 12-point Goudy.

Printed in the United States of America

October 2003

10 9 8 7 6 5 4 3 2 1

BVG

In memory of a cherished friend

Erik Y. Evren

1926–1999

who went ahead too soon

Contents

The Case of the Rhyming Robber

Police across the nation wondered: How did Idaville do it?

The town had sparkling white beaches, a Little League team, and a computer museum. It had churches, a synagogue, two delicatessens, and four banks. In short, Idaville looked like many other seaside towns.

But it wasn't.

Every person who broke the law in Idaville was caught.

How was this possible?

What was the secret?

Only Mr. and Mrs. Brown knew.

The mastermind behind Idaville's war on crime was their only child. They called him Leroy, and so did his teachers. Everyone else in Idaville called him Encyclopedia.

An encyclopedia is a book or set of books full of facts from A to Z, just like Encyclopedia's head. His friends

thought of him as a whole library that could whistle Beethoven.

Mr. Brown was chief of the Idaville police. He was smart and brave. His officers were well trained, honest, and loyal. But sometimes they came up against a crime they could not solve. Then Chief Brown knew where to go—home to dinner.

After saying grace, he went over the case.

Ten-year-old Encyclopedia listened carefully. When he had heard the facts, he asked one question.

One question was all he needed to solve a mystery.

Encyclopedia never spoke about the help he gave his father.

For his part, Chief Brown would have liked to announce to the world, "A bust of my son belongs in the Crimebusters' Hall of Fame."

But who would believe him? Who could believe that the mastermind behind Idaville's spotless police record was a fifth grader?

At dinner Tuesday, Chief Brown toyed with his soupspoon. Encyclopedia and his mother knew what *that* meant.

The police had come up against a case they couldn't solve.

"Do you want to talk about it, dear?" Mrs. Brown asked gently.

Chief Brown nodded. "A fortune in jewelry belonging to Mrs. Hubert Cushman was stolen from her home last week."

"Give Leroy the facts," Mrs. Brown said. "I'm sure he can help. He's never failed you yet."

Chief Brown laid down his spoon. "The thief who stole Mrs. Cushman's jewelry calls himself The Poet."

"I've heard of him," said Mrs. Brown.

"He steals jewelry and then sends a poem with a riddle in it to his victim," said Chief Brown. "The riddle tells where he buried the jewelry. Mrs. Cushman received her poem yesterday."

"How do his victims know if the riddle really tells where their jewelry is?" Mrs. Brown asked.

"He got careless twice," Chief Brown said. "He made the riddles too easy. The stolen jewelry was found."

"So he really does bury the jewelry," said Mrs. Brown. "What happens when the riddle isn't solved?"

"It's believed that he comes back sometime later, digs up the jewelry, and keeps it."

"My, is he ever something!" exclaimed Mrs. Brown.

"He's what is called a gentleman thief," explained Chief Brown. "Gentlemen thieves commit crimes mainly for the thrill. Outsmarting the police means more than the loot. It's all a sport with them."

Chief Brown took a piece of paper from his breast pocket. He unfolded it and handed it to Mrs. Brown. "This is the riddle Mrs. Cushman received."

Mrs. Brown read it, frowning.

"It doesn't make sense," she said. She passed the sheet

to Encyclopedia. "Here, Leroy. What do you make of it?"

Encyclopedia read The Poet's riddle:

Take the Landsmill Highway north,
And look along the border.
The second clue is marked in reverse,
But the first clue is in order.
The Poet

Encyclopedia had never been on the Landsmill Highway. Nevertheless, he closed his eyes. He always closed his eyes when he did his deepest thinking.

Mr. and Mrs. Brown waited anxiously.

A minute went by, and then another. Had the famous jewel thief, The Poet, outsmarted the boy detective?

Encyclopedia opened his eyes. He asked his one question. "Are there mile markers along the Landsmill Highway, Dad?"

Chief Brown was surprised by the question. "Why, yes, there are."

"Then," said the boy detective, "Mrs. Cushman's jewelry won't be hard to find."

Where was it buried?

(Turn to page 60 for the solution to The Case of the Rhyming Robber.)

The Case of the Miracle Pill

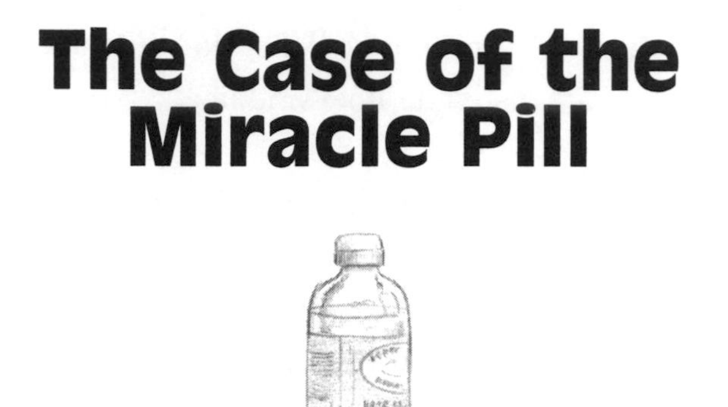

Encyclopedia helped his father all year round. During the summer he helped the children of the neighborhood as well. He opened his own detective agency in the garage.

Every morning he hung out his sign:

Brown Detective Agency
13 Rover Avenue
Leroy Brown, President
No Case Too Small
25¢ a Day Plus Expenses

To handle the tough kids, he took in a hard-punching fifth grader, Sally Kimball, as his junior partner. Sally was the prettiest girl in the fifth grade. She was also the best athlete.

One morning Encyclopedia and Sally had just opened the Brown Detective Agency for the day when Marsha Murphy stepped in.

"Take a look," she said. "This may be your last chance to see the old me. Soon I'll be in the money."

"Who says?" Sally asked.

"Wilford Wiggins," replied Marsha.

The detectives groaned.

A teenager, Wilford was as lazy as a time-out. Resting was what he did best. Whenever he got to his feet, he tried to fast-talk the little kids of the neighborhood out of their savings.

He never did. Encyclopedia always stopped his shady deals.

"Wilford has called a secret meeting for five o'clock today in the city dump," Marsha said. "He promised to make us little kids so rich we'll be the talk of the continent."

"What's he selling now," Encyclopedia asked, "a breakfast shake made of yeast and car polish for people who want to rise and shine?"

"Wilford's changed," Marsha said. "He told me so himself. He'll never tell another lie."

"Don't worry," Sally said. "You always know when Wilford is lying. His lips move."

Marsha's faith seemed to waver. She laid a quarter on the gas can beside Encyclopedia. "I want to hire you. Maybe Wilford isn't as honest as he says."

"We're hired," Sally said. "See you at the city dump at five o'clock."

When the detectives arrived, Wilford was standing behind a broken table.

On the table were an empty clear plastic bottle, an ice pick, a small jar, a drinking glass, and a pitcher filled with clear liquid.

Wilford started his big sales pitch.

"Gather around," he bellowed at the crowd of little kids waiting for him to fulfill their dreams of untold riches. They edged closer.

"Don't leak a word to any grown-up about the wonder I've got for you," he warned out of the side of his mouth. "They'll take over and cheat you out of every cent."

"Stop beating your gums and get to the big bucks," a boy shouted.

"You're keen for the green, eh, kid?" Wilford purred. "What I have for you today is Antiflow, the world's greatest gift to mankind! The savior of nations, the scientific marvel of the age! Remember the name: *Antiflow!*"

He unscrewed the cap on the plastic bottle and passed the bottle around. Next he took the ice pick and punched a tiny hole in the side of the bottle about an inch from the bottom. Then he filled the bottle from the pitcher. The liquid streamed out of the tiny hole.

Quickly he took a white pill from the jar. He held it up. "Observe: *Antiflow!*"

He dropped the pill into the bottle and screwed on the cap.

Although the bottle was still almost full, liquid stopped streaming out of the hole.

"Baloney!" a girl snapped. "It's a trick. There's something else in there."

"Oh, ye of little faith!" Wilford exclaimed. He filled the water glass from the pitcher and handed it to the girl. "Drink!"

She drank. "It's just water," she said, puzzled.

"Would Wiggins fool you?" Wilford cried. "The secret is the Antiflow. It was invented by Professor Stubblehauser of Germany. He doesn't trust anyone but straight shooters like yours truly. That's why he granted me the rights to sell the miracle pill in the U.S.A. He trusts me to give him half the profits."

Wilford paused to let the moneymaking possibilities of Antiflow sink in.

Then he said, "All my cash is tied up in oil wells. So I'm going to let my little friends in on this chance of a lifetime. For five dollars, you can buy a share in my Antiflow company. The more shares you buy, the more money you'll make!"

"Where's your factory?" a girl demanded.

"I'm glad you asked, friend," Wilford said. "I need your cash to help build the factory. When it's built, I'll make Antiflow by the ton. Don't miss out! Buy shares today at my special low-price, one-day-only offer."

The children talked excitedly among themselves. With Antiflow, floods would be a thing of the past. There were millions in it. Maybe more!

"Buy shares now," Wilford blared. "In a year you can afford to retire your mother and father."

That clinched it. The children lined up, eager to buy shares.

Encyclopedia hurried to the front of the line.

"Put away your money if you don't want a soaking," he said.

How did Encyclopedia know Antiflow was a fake?

(Turn to page 61 for the solution to The Case of the Miracle Pill.)

The Case of the Black Horse

Encyclopedia and Sally were straightening the Browns' garage when Waldo Emerson came in. He looked like he had stepped off a roof, or worse.

"Good to see you, Waldo," Sally said. "We haven't seen you round lately."

"Don't say that word!" Waldo howled.

"Sorry," Sally apologized. "I wasn't thinking."

Waldo had a thing about the word "round." Even when he heard it used harmlessly with other words, as in "round trip" or "round of golf," he threw a fit. It reminded him that some kids still believed the earth was round.

Waldo was the new president of the Idaville Junior Flat Earth Society. He was also the only member.

He laid a quarter on the gasoline can next to Encyclopedia. "I know the detective agency is closed until summer. But I want to hire you."

"What for?" Sally asked.

"I wrote an essay for Columbus Day tomorrow," Waldo answered. "The public library is giving a prize for the best essay about the explorer."

"What's the problem?" Encyclopedia asked.

Waldo moaned. "My essay was stolen yesterday. There isn't time to rewrite it. The contest closes at noon today. I wrote about how Columbus proved the earth was flat."

"How did he?" asked Sally.

"He didn't sail off the curve!" Waldo sang.

Encyclopedia never knew when Waldo was serious or having fun.

"I want you to get my essay back," Waldo said. "I'm sure Stinky Redmond stole it. He'll enter my essay as his and win the prize, a book called *The World of Dinosaurs*."

"Have you accused Stinky?" Sally asked.

Waldo rolled his eyes. "Yes, and of course he says he's innocent. He claims *he* wrote the essay. I dared him to meet me in half an hour in South Park at the carousel."

"Carousel," not "merry-go-*round*," Encyclopedia thought instantly. "Why at the carousel, Waldo?"

"The carousel is the scene of the crime," Waldo declared.

"But it doesn't open for an hour," Sally said.

"That's what I want," Waldo replied. "Stinky and I can have it out better with no one there to bother us. If Stinky doesn't show up, I'll know he's guilty."

On the bike ride to the carousel, Sally asked Waldo

what made him believe the earth was flat.

"Most of the earth is made up of water," he said, "and water is flat. Did you ever see a lake or a pond that had a hump in it?"

Sally and Encyclopedia admitted that they had not.

Waldo rolled on.

"The pictures taken of the earth from outer space are fakes," he insisted. "If the earth were a globe, China would be under the United States. My neighbor Mr. Chan comes from China. He would have hung by his feet when he lived there. He didn't. In fact, he has never hung by his feet in all his life."

Waldo reasoned like that.

Stinky Redmond was waiting by the carousel. The double ring of ride-on animals stood silent and still.

Waldo sneered at Stinky. "The thief returns to the scene of his crime! He thought he could get away with stealing my essay!"

"Let's hear what happened," Encyclopedia suggested.

Waldo said, "When I climbed onto the carousel platform yesterday, Stinky was already standing by that black horse."

He pointed to a black horse with one hoof off the ground. It seemed about to prance away. Like all the animals, it had a pole through the back of its neck.

Waldo pointed to a bench beside the black horse. "Before the carousel started, I laid my bag with the essay on that bench," he said. "Stinky swiped it."

"I didn't swipe his bag!" Stinky broke in. "He's afraid my essay will beat his. Mine is oh, so funny. It'll win in a laugh. I wrote that Columbus proved the earth is flat!"

Waldo harrumphed and continued. "Then I got on that white horse," he said, pointing to a white horse three horses in front of the black one. "I never saw you get on the black horse. I never saw you get off. But when the ride ended, I did see you run from the bench like you were legging it for a lifeboat."

"This kid isn't two days out of his tree," Stinky growled. "Sometimes I get sick going up and down, even on a seesaw."

"You weren't on a seesaw," Waldo snapped.

Stinky retorted, "I started getting sick when the black horse moved up and down on the pole as the carousel turned. So I got off and sat on the bench until the ride ended and I felt better. There was a bag on the bench, but I never touched it."

"Why did you rush off the carousel when the ride ended?" Sally asked.

"I had to go to the you-know-where," Stinky mumbled.

Sally whispered to Encyclopedia, "I don't know who to believe. Maybe Waldo never wrote an essay and he's trying to get Stinky in trouble by saying he's a thief. Or maybe it's Stinky who never wrote an essay and stole Waldo's."

"Take another look at the bench and the horses," Encyclopedia suggested.

"I'm looking," Sally said. "Stinky's black horse is three

horses directly behind Waldo's white horse. The bench is just to the left of the black horse."

"Is that all?"

"Yes, except I wish the horses could talk and tell us who is lying."

"One has, in its way," replied the detective.

Who lied, Stinky or Waldo?

(Turn to page 62 for the solution to The Case of the Black Horse.)

The Case of Nemo's Tuba

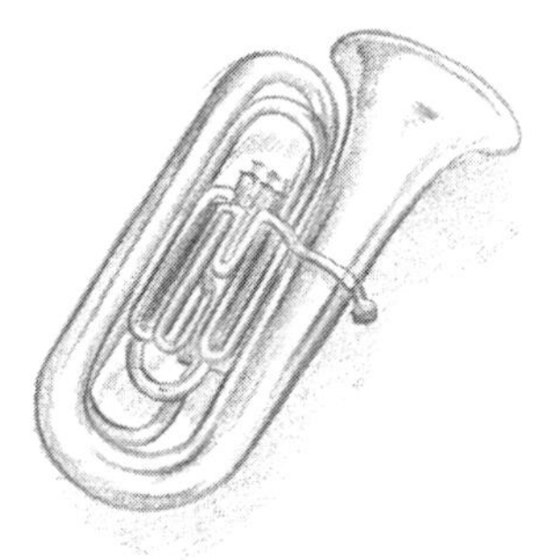

The detectives were closing the agency for the day when Nemo Huffenwiz, a pudgy sixth grader, blew in. He plunked twenty-five cents on the gas can.

"What's on your mind?" Encyclopedia asked.

"My tuba," Nemo announced. "I know what you're thinking. The tuba is for Tubby Tuba, the fat kid in the back row of the school orchestra."

"Anyone who calls you Tubby Tuba should have his valves ground off," Sally said.

"How may we be of help?" Encyclopedia inquired.

"Find out who played a dirty trick on me and my tuba," said Nemo.

He went over the details of the case for the detectives.

That afternoon the summer youth orchestra had given a performance of Suchalicki's "March of the Frosty Flowers"

in the school auditorium. Nemo was delayed at the dentist's office and arrived at the school late.

Grabbing his tuba from the music room, he had raced to his seat in the rear of the orchestra. Mr. Downing, the conductor, had just raised his baton.

"Boy," said Nemo, "did he give me a look. It curled my shoelaces. Lucky for me I didn't have to play for a while."

"Is it true that in many pieces the tuba doesn't play a single note?" Sally asked.

"Yep," replied Nemo. "In 'March of the Frosty Flowers' the tuba plays only one note. I sweated out forty-two measures before I played it, a high E. If I hit it, I was a hero. If I missed, I was a bum."

Sally cried, "Tell us!"

"I was a bum."

"Oh, dear."

"Someone switched the valves on my tuba," Nemo said. "The valves are what you push down to make the sounds. They should be in order, one, two, three. Someone switched them to three, one, two. You can't tell just by looking if they're in the right order or not. You have to blow."

"Who could have switched them?" Sally asked.

"Anyone," Nemo said. "The instruments belong to the school and are kept in the music room. Kids can practice there or take the instruments home."

"Lugging a tuba home will flatten your feet flatter than a flatiron," Sally warned.

"You're so right," replied Nemo. "That's why the school doesn't allow students to take home the two bass fiddles or the tuba. I practiced the tuba yesterday until the janitor locked the music room for the night."

"Do you suspect anyone?" Sally asked.

"Alma Higgens," Nemo said instantly. "The instruments were handed out in the fall. I got the tuba. Big Alma had wanted it so she could show that a girl is strong enough to carry it. But I was there a minute before her. She had to settle for a trumpet."

"That shouldn't have made her mad enough to pull such a dirty trick on you," Sally said.

"There's more," Nemo said. "This morning I was throwing a baseball with Mitch Jennings on his front lawn. Alma rode by on her bike. I threw wide. The ball whacked her on the foot and she fell off her bike. Oh, boy, was she mad."

"It isn't a good idea to make Alma mad," Sally said. "She's hotheaded."

"The throw was an accident," Nemo said. "But the fall hurt her. Her lip was cut and bleeding a bit, and she was limping. I was sorry and tried to apologize. You should have heard what she said to me!"

"Alma could have sneaked into school before the performance of 'March of the Frosty Flowers' and switched the valves on your tuba," Sally said. "Limping or not."

"I believe it's time to question Alma," Encyclopedia said.

On the way to Alma's house, he stopped at the school to speak with Mr. Downing, the conductor.

When Encyclopedia came out, he said, "Mr. Downing told me Alma telephoned him before the concert. She said she hurt herself falling off her bike this morning. She was staying home to rest and practice the trumpet."

At Alma's house, Sally rang the doorbell. Alma opened the door.

"What's the squawk?" she demanded, glaring at Nemo.

"I just remembered," Nemo whispered to Encyclopedia. "I don't want to be here."

Sally stepped fearlessly up to the big girl. "We think you switched the valves on Nemo's tuba. You wanted him to miss his one big note, a high E."

"So Mr. Downing would think I wasn't good enough and kick me out of the orchestra," Nemo put in, his courage up. "Then you'd take over the tuba."

"Think again, mousehead," Alma snarled. "I wasn't near the school today."

"Where were you?" Sally demanded.

"After I telephoned Mr. Downing to tell him I couldn't make it today because I hurt my foot, I went to my room. I read and practiced the trumpet."

"How come we didn't hear you playing?" Sally demanded.

"I quit practicing a few minutes ago," Alma said. "Besides, I use a mute. I can barely be heard in the next room. I'm the kind who respects the ears of others. Now bye-bye, you sand fleas."

She shut the front door with a bang.

"I sure hope she's guilty," declared Nemo. "If she isn't, she's going to make me pay for saying she is. Maybe I ought to give her the tuba and take up barrel jumping."

"Don't," Encyclopedia said. "Alma switched the valves."

How did Encyclopedia know?

(Turn to page 63 for the solution to The Case of Nemo's Tuba.)

The Case of the Ring in the Reef

Hector Heywood was nearly in tears when he came into the Brown Detective Agency.

"Bugs Meany, that no-good bully!" he wailed.

"Oh, not Bugs again!" exclaimed Sally.

Bugs was the leader of a gang of tough older boys. When Encyclopedia and Sally weren't around, they bullied the little kids of the neighborhood.

The boys called themselves the Tigers, but they should have called themselves the Spurs. They always arrived on the heels of trouble.

Sally often said Bugs was quite accomplished for a boy with the IQ of a refrigerator door.

"What's Bugs done now?" Encyclopedia asked Hector.

"He stole Mrs. van Colling's diamond ring from me," Hector replied.

He explained. He had been at the beach the day before. He'd found a ring in the sand. The ring hadn't looked valuable, but he had taken it home.

"This morning the *Idaville Gazette* had a story about the ring," Hector said. "It's worth a lot of money, and there's a reward for finding it. The newspaper said the ring belonged to Mrs. van Colling."

"I read the story," Encyclopedia said. "Mrs. van Colling thought she had lost it while scuba diving at Warren Reef. She hired two divers to search for the ring. They didn't find it."

"That's because she lost it on the beach," Hector said. "I was returning the ring when Bugs and three of his Tigers stopped me a block from her house. They asked where I was going, and like a dummy I told them. They turned me upside down. They shook me until the ring fell out of my pocket."

Hector laid a quarter on the gas can. "Get the ring back. I found it. I should get the reward, not Bugs."

Encyclopedia agreed. "We'll go see Bugs."

"You go," said Hector. "I'm about to do what any red-blooded coward would do—go home. Bugs is too rough for kids our age."

"Except one," Encyclopedia said. "Sally has straightened him out before."

It was true. The last time Sally and Bugs had fought, the toughest Tiger had taken a right to the nose. His eyes had

rolled up far enough to see his brains. For a full minute he had staggered around as if looking for the rest of himself.

"Okay, I'll go with you to see Bugs," Hector said. "But you'd better have an escape plan."

The Tigers' clubhouse was an unused toolshed behind Sweeney's auto body shop. Bugs was sitting on a crate out front.

He had a deck of cards and was practicing dealing himself all the aces.

When he spied the detectives and Hector, he growled, "Well, well, the little goody-goods." His lips curved in a sneer. "Go adopt an egg!"

"Don't get your dandruff up, Bugs," Sally said. "Hector told us he found Mrs. van Colling's ring on the beach yesterday. He says you took it from him."

"What is this?" Bugs growled. "You dare accuse Bugs Meany, the idol of America's youth, of being a common thief? I found the ring! I'm waiting until I think Mrs. van Colling has had her breakfast before I return it. I'm a gentleman."

He took a step toward Hector, his teeth bared.

"I don't think he wants to be friends," Hector whispered to Encyclopedia. "I have what I believe is a very good idea: Run for your life!"

Encyclopedia grabbed his arm and held him.

"Stay calm," the detective said. "Trust Sally."

"Where did you find the ring, Bugs?" Sally demanded.

"I often dive at dawn," Bugs purred. "The reef is so beautiful then! The pursuit of beauty is my life. I don't get along on good looks alone."

"Where did you find the ring?" Sally repeated.

"If you must know, I was swimming by the reef when my foot struck something lying on the bottom," Bugs said. "It was a bright yellow fish, dead. I moved it with my foot. I saw what had been lying under it. A ring!"

"Aw, c'mon, Bugs," Sally said. "That's the biggest fish story I ever heard."

"You doubt the word of Bugs Meany?" Bugs said, his voice rising. "Nobody gets away with calling me a liar!"

He took aim and threw his Sunday punch. Sally sidestepped and cracked him one on the side of the jaw. *Zowie!*

Bugs spun like a propeller. Encyclopedia thought he saw Bugs's face and the back of his head at the same time.

Bugs slowed, wobbled, and fell flat. He lifted his head and moaned, "I hate it when she does that."

Sally suddenly looked concerned. "Oh, no. Maybe he's telling the truth!"

"He isn't," Encyclopedia said.

What made Encyclopedia so sure?

(Turn to page 64 for the solution to The Case of the Ring in the Reef.)

The Case of the Lawn Mower Races

Encyclopedia and Sally arrived at the county lawn mower racing championships at five minutes to nine on Saturday morning.

Engines roared. Billows of dust and smoke rose from the track, a nine-acre pasture on Josh Woodly's farm. Rows of spotlights and bales of hay outlined the course.

Souped-up, bladeless riding mowers bounced over ruts and bumps at speeds of more than forty miles an hour. The riders were finishing the last few laps of the twelve-hour endurance race, which had begun at nine o'clock the night before.

Three of the detectives' classmates—Larry Winslow, Bill Marshall, and Ken Uster—were in the crowd.

"You got here just in time," Larry said to the detectives. "This race ends in a few minutes."

Encyclopedia watched the seven mowers scream around the curves.

"How come there are only seven mowers in this race?" Sally asked.

Ken explained. Twenty mowers had started, each driven by a team of two men and a woman. Fifteen teams had to retire. Most of them had gone to the hospital with headaches, sprains, cuts, and bruises.

"Racing lawn mowers can be hard on your health," Bill observed.

The crowd began cheering. A teenager, Mary Mullins, had taken the lead. She crossed the finish line the winner.

Her two teammates rushed to congratulate her. Autograph hunters held out pen and pencil. She looked very tired, but she signed everything put in front of her.

"The next race is a one-miler," Larry said. "The rules are different from the twelve-hour race. No pit stops for fuel are allowed. The engine has to be the one that came with the mower. So the mowers can't go as fast as in the twelve-hour race."

"What you *can* do," added Bill, "is change the driving gears or pulleys to send more power to the wheels. That gives you more speed."

The contestants in the one-mile race were announced. Larry's twin brother, Bill's aunt, and Ken's cousin were all competing. Mary Mullins was in the race, too. The

detectives and a few of the crowd, including Bill, Ken, and Larry, followed Mary to her trailer in the parking lot. On the trailer was her shiny new mower. It had yellow racing stripes and countless coats of polish.

She tightened the wide belt that held her insides in place going over the bumps. After adjusting her crash helmet, she pushed the mower to the starting line.

The command came. "Ladies and gentlemen, start your engines!"

All ten mowers sputtered to life.

The drivers leaped aboard. They quickly attached a cut-off switch to their bodysuits. If a driver was thrown off the mower, the engine stopped.

The starting gun sounded. The mowers roared down the course.

Mary Mullins didn't get far. She hit two bumps and her mower died.

Jumping to the ground, she peered at the engine. Ken, Larry, Bill, and the detectives rushed over to help.

"The nut and bolt that hold the line from the gas tank to the carburetor fell off," she said. "Luckily, the line hung upright or gas would have spilled all over the course. I just lost the gas in the carburetor."

She shook her head, puzzled. "Everything was tight yesterday. The engine ran beautifully."

She glanced at the mowers racing away. "If I find the

nut and bolt and can get wrenches quickly, I can at least finish."

"I'll get the wrenches," Bill volunteered. "Where do you keep them?"

"All my socket wrenches and open-end wrenches are in the toolbox in my truck," Mary said.

"I'm on my way," Bill said.

While he was gone, Ken, Larry, and the detectives helped Mary search for the nut and bolt.

"Here's a nut!" Ken cried. He picked it up from the grass a few yards past the first bump. Larry found a bolt past the second bump.

Mary said the nut and bolt were hers.

Bill came running back. "I brought an open-end wrench *and* a socket wrench just to be safe."

Using the wrenches, Mary tightened the bolt and nut in place. She started the engine and set off in hot pursuit of the other mowers.

"The nut and bolt were loosened by someone who didn't want Mary Mullins to win the one-mile race," Sally mused. "It was easy to do while everyone was watching the finish of the twelve-hour race."

"That makes sense," Encyclopedia murmured.

Sally frowned. "The guilty person wanted someone besides Mary to win. That makes Larry, Bill, and Ken suspects. They each had a family member in the race. Or it could have been someone else. *Anyone!*"

"No, not just anyone," corrected Encyclopedia.

"Don't tell me you know who did it," Sally said.

Encyclopedia nodded. "I do."

Do you?

(Turn to page 65 for the solution to The Case of the Lawn Mower Races.)

The Case of the Jumping Frogs

Buddy Mayfair, better known as Ribbet, was the only fifth grader anyone knew who ran a college for frogs.

"It's frog-catching season," he announced, hopping into the Brown Detective Agency.

During frog-hunting season, Ribbet warmed up by hopping a lot.

"The science club is hunting frogs at South Park at two o'clock today," he said. "Stinky Redmond and Alma Higgens aren't coming. Want to take their places?"

Encyclopedia and Sally didn't have to be asked twice. They had never been on a frog hunt.

"I'm glad Stinky and Alma aren't coming," Sally said. "They're always a problem."

A high steel fence surrounded the campgrounds at

South Park. Standing by one of the two gates was a park ranger. Beside him were the other campers and Mr. Sands, the science teacher. Mr. Sands was in charge of the outing.

He welcomed the detectives, who were the last children to arrive. "We'll be the only group in the campgrounds today. Come inside."

The ranger locked the gate behind them. He climbed into a small pickup truck and drove off.

"There are two gates, and both are locked when everyone is inside the campgrounds," Ribbet said. "It's for the campers' safety. Mr. Sands has a key—"

He broke off. A rumbling had grown into a full roar. A huge tractor was coming straight toward the gate.

The driver unlocked the gate, drove in, and then locked the gate behind him.

"I'm cutting a firebreak," he told the children. "Sorry about the noise the disks make."

Behind the tractor was a row of twelve steel cutting disks. At both ends of the row were wheels with thick tires. The wheels could be lowered, thus lifting the disks off the ground when not in use.

With a friendly wave to the children, he slowly drove on. The disks churned roots, shrubs, and stumps, leaving a fireproof trail.

Unlocking the far gate, he drove through and locked it again.

“Frogs are waiting,” Mr. Sands called when the campers had pitched their tents and spread their bedrolls. “Let’s head to the pond.”

The boys and girls hurried through the gate, which Mr. Sands locked behind them.

“You can’t be too careful,” Ribbet said. “Lots of things have been stolen from the park lately.”

The pond lay beyond a woods out of sight of the campgrounds. During the walk, Ribbet fine-tuned himself with hops every few steps.

“I hope the hunting is good,” he said. “I need a big enrollment.”

For the past two years Ribbet had trained frogs for others at his frog college. His students always did well at frog-jumping contests.

Not only did Ribbet teach his student frogs quick takeoffs and how to race. They got a room and all the flies they could eat.

Other services included massages and calisthenic drills, plus time off to swim in the family birdbath.

All for fifteen cents a day.

“I started the college two years ago after my bullfrog, General Grant, set a state record,” Ribbet said. “The General covered seventeen feet two inches in the three jumps. I retired him. He’s now my poster boy.”

“How’s business this year?” Encyclopedia inquired.

“Only fair,” Ribbet answered. “Right now I’m training frogs for two grown-ups and three kids.”

The day's hunting didn't go well. Although the *ribbet, ribbet* of frogs filled the air, the only creatures found were tadpoles.

"We'll do better tonight," Encyclopedia said.

Back at the campgrounds the children perked up by playing among the tracks in the firebreak. Led by Ribbet, they leapfrogged from one tire track to another. He was having so much fun he ran to get his camera from his tent.

The fun ended when Ribbet discovered that his camera had been stolen.

"I should have taken it with me to the pond," he groaned. "But I was afraid of dropping it in the drink."

"The thief can't be one of us," Sally insisted. "The whole group went to the pond together and came back together."

Mr. Sands was upset by the theft. He fetched the ranger.

The ranger examined the locks on both gates. They were unbroken.

"Only you and me and Hal, the tractor driver, have a key to the gates," the ranger said to Mr. Sands. "Hal said he never returned to the campgrounds after cutting the firebreak. I stayed in my office after I left you."

The ranger chuckled. "I don't think it was a thief at all," he said. "Probably it was a caterpillar. Everyone knows what shutterbugs they are. The camera will turn up."

"Only a fool of a thief would risk climbing over the fence," Mr. Sands mused.

"The thief didn't have to," said Encyclopedia.

Mr. Sands stared at Encyclopedia, startled. "Do you know who stole the camera?"

"Why, yes," said the detective, "though he had me stumped for a time."

Who was the thief?

(Turn to page 66 for the solution to The Case of the Jumping Frogs.)

The Case of the Toy Locomotive

On Wednesday Encyclopedia and Sally went with Sol Calvin to the Best Buy Toy Company factory for the yearly auction. Sol's sister Birdie worked there.

She tested toys, though she didn't know it.

Birdie was four years old.

The toy factory was in a large red brick building. Sol led the detectives to a room in which there were five small children aged three to four, two teachers, and lots of toys. One of the teachers was reading a book of fairy tales to the children.

"Golly," Sally said. "It looks like a classroom in a nursery school."

"It is," Sol said. "The little kids report for classes here three days a week. They listen to stories, finger paint, dress up in costumes, and eat snacks. The most important period is free time. That's when they play with the toys the company makes."

"What toys are being sold today at the auction?" Sally asked.

"Toys the kids are tired of," Sol replied, "or toys that have flopped. If the kids don't like a new toy, it doesn't get into stores."

"Is that good business?" objected Sally. "I mean, five little kids forcing their tastes on the whole country."

"The Best Buy Toy Company doesn't let that happen," Sol answered. "The testers change. A new group of little kids enrolls every two weeks."

He pointed to a large mirror.

"It's a one-way mirror," he said. "Company officers sit behind the mirror. They can see the kids but the kids can't see them."

"The kids have no idea they're testing the toys?" Sally said.

"Correct," Sol said. "The officers and the teachers make careful notes about which new toys are liked or disliked."

"That's neat," said Encyclopedia. "The kids don't have to give answers they feel the company wants to hear."

The teacher who had been reading to the class closed the book. "Free play time," she announced.

The young toy-testers immediately went to work. Some chose toys and played quietly and happily.

Others weren't happy. They had chosen one or another

of the newer toys. When it failed to please, it was shaken, sat on, or cast aside.

Sol's sister Birdie was busily banging a red car on the floor. A wheel flew off.

Birdie wailed.

"That car will go back to the drawing board," Sol said.

The auction was going to start soon. The small toys were displayed on a long table. Bigger toys were on the floor. An electric locomotive caught Encyclopedia's eye.

"Do kids this little really play with electric trains?" he asked.

"Hardly," Sol said. "The locomotive belongs to one of the men in bookkeeping. It's pretty beat up, but he hopes it will sell. I want it. If the bidding goes past five dollars and twenty cents, though, I'm sunk. Five dollars and twenty cents is all I have."

The nursery was filling up with parents. Just before the bidding began, Sledge O'Hara, Bugs Meany's eleven-year-old cousin, came in.

Sledge's right arm was in a sling. He had hurt his shoulder badly in a card game trying to catch a joker. It had fallen out of his sleeve.

The auction began. The prices were a bargain hunter's dream. Everyone was having fun except Sol.

His bid for the locomotive fell eighty cents short. Sledge bought it for six dollars.

"What happens if the locomotive doesn't work?" Sledge

demanded of Mr. Wilmott, the auctioneer and a vice president of the Best Buy Toy Company.

"Return it, and the company will let you pick another toy free of charge," said Mr. Wilmott. He put the locomotive in a gift bag.

Sledge slipped his left arm through the handles and slung the bag over his left shoulder. He strutted from the nursery, grinning slyly.

Ten minutes later he was back, empty-handed.

"Two big teenagers stole my locomotive!" he howled.

He had left by the revolving door in the rear of the factory, he said. A big teenager got in the slot ahead of him and dropped a package. It jammed the door.

"I was stuck, trapped like a rat!" Sledge cried.

Another big teenager, he went on, snatched the bag with the locomotive. Both thieves got clean away.

"I was robbed in your building," Sledge said to Mr. Wilmott. "You're responsible. You owe me for my pain and suffering. But I'm not pushy. I'll take another toy."

"Sledge is such a liar and a cheat," Sol muttered. "I'd go over Niagara Falls on a banana peel before I'd believe him."

"Did anyone see the thieves?" Mr. Wilmott asked calmly.

"No, the theft is just another chapter in my life of toil and hardship," Sledge moaned.

"Why did you leave by the back door?" Mr. Wilmott inquired.

"Because my bus home stops behind the building," Sledge answered. "The streets are so unsafe these days, I wanted to get to the bus as quickly as possible. An honest lad like me doesn't stand a chance alone. We're in a crime wave! Bad guys are everywhere!"

Mr. Wilmott hesitated, and then said, "Well . . . all right. Pick another toy."

Sledge grinned and tapped a music box. "This will do," he sang.

"Encyclopedia!" Sally exclaimed. "Sledge will walk off with both the locomotive and the music box for six dollars! You can prove he wasn't robbed, can't you?"

The detective smiled his knowing smile. "Of course."

What was the proof?

(Turn to page 67 for the solution to The Case of the Toy Locomotive.)

The Case of the Air Guitar

Encyclopedia and Sally were biking in town when they saw Scott Burlow in an alley. Scott was dancing like a chicken on a hot stove.

The fingers of his left hand, which were next to his shoulder, slid up and down, twitching like crazy. His right hand seemed to be scratching his hip, and he was shaking his head so fast that it looked like it might snap right off his neck.

"Oh, dear," said Sally. "I hope he doesn't get whiplash."

"Maybe he just ate his first raw oyster," Encyclopedia offered hopefully.

Scott saw the detectives and stopped dancing and twitching and scratching and shaking.

"Scott! What itches?" Sally inquired anxiously.

Scott laughed. "Nothing. I'm tuning up is all. Today is the day."

"For what?" Encyclopedia was almost afraid to ask.

"You're detectives," Scott said. "I'll give you a clue. Check out my hair."

"It's long," Sally said.

"I grew it an extra five inches for the air guitar contest," Scott said. "You have to look like a musician if you want to catch the eye of the judges."

Encyclopedia had heard of air guitar contests, in which the performers pretended to play a guitar. The guitar couldn't be seen or heard because it didn't exist.

"I finished third last year," Scott said. "I'm not resting on my laurels. A win today and I'm in the state finals."

Sally said, "We were on our way to the early movie. But we'd rather see you play air guitar."

"Come on. The contest starts at eleven-thirty at the dance school on Third Street," Scott said.

The detectives went with Scott to the dance school. The main room had some two dozen folding chairs in front of a stage.

The stage was empty except for a piano.

"I'll show you around," Scott said.

He led the detectives backstage, where the equipment was stored.

He pointed toward a door. "That leads to the office."

"There's someone in there," Sally whispered. "Listen."

Two boys were speaking in low voices.

"Make sure it's his and not one of ours."

"It's his. Let's go. It's nearly twenty minutes past eleven."

"You're fast. It's only eleven-eighteen—oops, eleven-nineteen."

"Who's in there?" Scott called.

Encyclopedia opened the door as the other door in the office slammed shut. He crossed the office and opened the other door. It led to the street.

"We're too late," Encyclopedia said.

Whoever had been in the office had turned the corner and was out of sight.

"I don't like this," Scott muttered.

They went back into the main room and took seats. Friends and relatives of the air guitar players were filing in.

Mrs. Watson, the elementary school music teacher, sat down at the piano. She placed several sets of sheet music on the rack above the keyboard.

"Each person performs to the music he's chosen," Scott explained.

At eleven-thirty, Mr. Jurgens, one of the two judges, announced the six contestants.

Scott and the other five performers took off their wrist-watches and laid them on the piano.

"Each kid has exactly one minute to perform," Scott explained upon returning to his seat beside the detectives. "You lose points if you stop playing more than ten seconds before or after the music ends."

"So no one can cheat by checking his watch?" said Sally.

"That's the idea," Scott said. "The judges look for artistic style, ability to stay with the music, and airiness."

First up was Adam Lang. He wore a red wig and played his invisible guitar to "Rowdy Rob Robin."

He flung himself this way and that. Alas, he became dizzy, lost his footing, and knocked himself out in twenty-two seconds.

Scott, Harold Johnson, Phil Twining, and Manny Foster had their turns. Each strummed the air wildly, hopping and flopping as though hooked to a live wire.

"It's going to be a tough call," Scott remarked. "We all cracked down without cracking up."

Herb Carter was last to perform. "Herb won last year," Scott said. "He really rattles his bones. He's going to be hard to beat."

Herb swaggered onto the stage, grinning confidently. He spread his feet, ready to do his thing the moment Mrs. Watson started playing.

She didn't play. She fumbled through the stack of music she had on the piano.

"I can't find Herb's piece!" she exclaimed. "I'm sure I had it in the office with the other music for safekeeping."

"Those two boys who were talking in the office stole Herb's music!" Sally whispered. "It has to be! But which two?"

"Couldn't you tell by their voices who the thieves are, Scott?" Encyclopedia asked.

Scott shook his head. "They were speaking too quietly,"

he answered. "Without his own music, Herb doesn't stand a chance."

Herb performed bravely, but clearly not well enough to win.

The judges counted up the scores.

Encyclopedia used the break to stroll onto the stage and over to the piano. Sally followed him.

The boy detective studied the watches.

Five of the watches had a minute hand and an hour hand. The sixth watch had no hands—it showed the time digitally.

"*Hmm,*" Encyclopedia said.

"What does '*hmm*' mean?" Sally demanded.

"It means I know who one of the thieves is," Encyclopedia answered. "From him, we'll learn the name of the other."

What was the clue?

(Turn to page 68 for the solution to The Case of the Air Guitar.)

The Case of the Backwards Runner

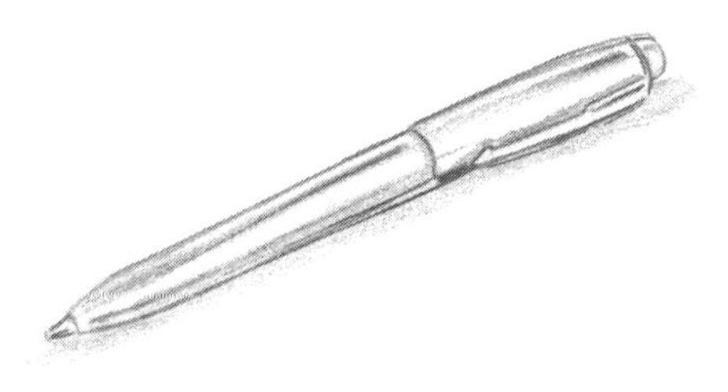

Encyclopedia and Sally were biking by the Grove Shopping Center when they saw a fight about to break out.

Felix McGee and Rupert Dugan were on the sidewalk near the exit lane of the parking lot, screaming in each other's faces.

Oscar, a security guard, was trying to keep them from putting their fists in each other's faces.

It wasn't easy. Both boys were built like barrels.

Felix played tackle on the seventh-grade football team. Rupert was the star of the seventh-grade wrestling team.

"Those two have never liked each other," Sally said. "In fact, they hate each other."

"We'd better find out what this is about," Encyclopedia said.

The detectives walked their bikes onto the sidewalk.

Felix shook his fist under Rupert's nose. "You're cleared for takeoff, fatso!"

Rupert shook his fist under Felix's jaw. "If I hit you with this, you'd better have wings that fit you."

Oscar was glad to see Encyclopedia and Sally. He wouldn't have to separate the two boys by force. He let the young detectives take over.

Felix and Rupert were glad, too. They were saved from having to make good their threats to punch each other out.

Oscar told the detectives what had happened. A few minutes earlier, he had seen a boy palm a silver pen from a counter in Fabian's Gift Shoppe. The boy shoved it into his pocket and legged it out of there.

"I chased him outside," Oscar said. "He ran from the parking lot down the exit lane to the street."

"You never saw the thief's face?" Sally asked.

"No, I saw only his back," Oscar said. "I lost sight of him for a moment when a pickup truck drove between us. Then I saw these two boys. Both were dressed like the thief, in a white T-shirt, blue shorts, and sneakers. They were quarreling. Neither had the pen on him."

"That's because I saw Felix toss something small into a pickup truck that was driving past," Rupert said. "It must have been the pen he stole."

"He could have," Oscar admitted. "I didn't see that because the pickup truck passed between us."

"Felix had to get rid of it before you caught up with

him," Rupert continued. "It was evidence. Now it's out. He's been living a dark and secret life—the life of a shoplifter!"

"I didn't steal anything. I didn't toss anything into a truck," Felix insisted. "Rupert did."

"Then why were you running?" Encyclopedia asked.

"Because I had to be home by noon for my aunt's birthday party. I ran because I realized it was noon and I was late."

"Without a watch, how did you know it was noon?" Sally said, pointing to Felix's bare wrist.

"I was in the bookstore when I heard the bells in the church behind the shopping center. They chime at noon," Felix replied.

Oscar nodded. "The church bells did chime."

Felix said, "My house is only three blocks away. I figured if I ran like crazy I wouldn't be more than a couple of minutes late. Now I'll really be late."

Encyclopedia turned to Rupert. "What were you doing near the shopping center?"

"I was jogging," he answered. "I like to keep in shape."

"The shape of a watermelon," Felix said.

"I had jogged past the exit lane when I saw Felix running down it," Rupert said. "I saw him toss something small and shiny into a passing truck. It must have been the pen. I stopped to see what was going on."

"How could you see Felix running down the exit lane?" demanded Sally. "You just said you had jogged *past* it. That means you had your back to him."

"I was jogging backwards when I saw Felix running," Rupert said. "I always jog backwards when the sun is in my eyes, like today. Everything that forwards running messes up, backwards running puts right. Backwards running helps the knees and hips and is easier on the joints. But it's tiring."

"You hunk of blubber!" Felix cried. "You didn't see anything."

"I know what I saw," Rupert said. "You didn't want Oscar to find the pen on you, so you threw it into a passing truck. My eyes don't lie."

"Oh, yeah?" Felix snapped. "You remind me of an ostrich. An ostrich's eye is bigger than its brain."

"Oh, yeah?" Rupert retorted. "You remind me of a starfish. A starfish *has* no brain."

"Encyclopedia," Sally whispered. "What do you think? Felix could have stolen the pen and tossed it into the truck before Oscar caught up with him. Or Rupert could have stolen the pen and dumped it in the truck himself."

"A case about a pen is a case about words," answered the detective. "Therefore, the boy who didn't tell the truth is—"

Who didn't tell the truth, Rupert or Felix?

(Turn to page 69 for the solution to The Case of the Backwards Runner.)

Solutions

SOLUTION TO

The Case of the Rhyming Robber

When Chief Brown said there were mile markers along the border of the Landsmill Highway, Encyclopedia knew where the jewelry was buried.

The last two lines of the riddle told him.

The line "But the first clue is in order" meant that the first number of the mile marker was forty. It is the only number in the English language whose letters are in alphabetical order.

The second number of the mile marker was one. It is the only number in English whose letters are in reverse alphabetical order.

Therefore, Mrs. Cushman's jewelry was buried by or under marker forty-one.

Chief Brown ordered a stakeout of mile marker forty-one. Six days later, The Poet was captured as he dug up Mrs. Cushman's jewelry.

SOLUTION TO

The Case of the Miracle Pill

Encyclopedia realized what was stopping the water from coming out of the small hole near the bottom of the bottle.

It was not the Antiflow pill, which was nothing but a piece of wood painted white.

It was the bottle cap.

Had Wilford not screwed the cap on tightly, the water would have continued to come out of the hole.

Prove this for yourself. Do as Wilford did. Let water flow out a small hole near the bottom of a plastic bottle.

Now press the palm of your hand over the top of the bottle or screw the cap on tightly.

The water will stop coming out until you remove your hand or loosen the cap!

Wilford was forced to admit the pill was a fake and stopped trying to sell Antiflow.

SOLUTION TO

The Case of the Black Horse

Stinky said he had become sick when his horse moved up and down on its pole. He had gone to the bench to recover.

He had already seen Waldo carrying the bag toward the carousel. He had stood by the black horse as if preparing to mount. The black horse was three horses behind Waldo's white one. Hence, once the carousel began to turn, Waldo was unlikely to look back and see what Stinky was up to.

Stinky never sat on the black horse.

When the ride started, he went straight to the bench to see what was in the bag. He was too busy reading Waldo's essay to notice his mistake.

But Encyclopedia noticed.

A carousel horse with three feet (or four feet) on the ground, like the black horse, doesn't move up and down. It doesn't move at all! Stinky couldn't have gotten sick!

Stinky returned Waldo's essay. It won second prize, a globe.

SOLUTION TO

The Case of Nemo's Tuba

Alma wanted to get even with Nemo for beating her to the tuba and knocking her off her bike. So she telephoned Mr. Downing, the orchestra conductor. She told him that she was hurt and couldn't get to school to play in "March of the Frosty Flowers." She promised to practice the trumpet at home.

Alma thought that gave her an alibi. How could she have switched the valves on the tuba if she wasn't at the school?

Encyclopedia knew better.

She had hurt not only her foot, but her lip, too. It was bleeding when she fell off the bike.

She couldn't have been at home blowing a trumpet with a cut lip.

Caught in the lie, she confessed. She had switched the valves on Nemo's tuba.

Mr. Downing moved her from the trumpet to the triangle.

Nemo kept the tuba.

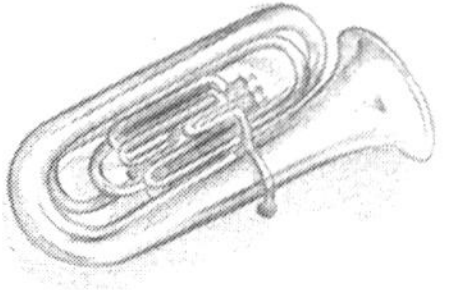

SOLUTION TO

The Case of the Ring in the Reef

Bugs had also read about the lost ring in the *Idaville Gazette*. He had to say he found the ring in the water around the reef where Mrs. van Colling thought she had lost it.

That would prove how smart he was, smarter than the divers she had hired.

The divers would have searched all the parts of the reef. But they wouldn't have bothered to look under a dead yellow fish lying on the bottom. At least that's what Bugs thought.

He was right. The divers didn't look under the fish because it wasn't there.

Fish that haven't been dead long enough to lose their color don't sink to the bottom. They float near the surface.

Caught like a fish out of water, Bugs admitted taking the ring from Hector. Truth, he decided, hurt less than Sally's fists. He gave back the ring.

Hector received the reward, a wristwatch that told time underwater.

SOLUTION TO

The Case of the Lawn Mower Races

Having just finished the twelve-hour race, Mary Mullins was very tired. So she accepted Bill Marshall's offer to go and get wrenches rather than get them herself.

Bill offered to get the wrenches to throw suspicion off himself by helping Mary.

He didn't fool Encyclopedia.

Being tired, Mary Mullins forgot to tell Bill the size of the wrenches she needed. But he fetched the right size wrenches for the nut and bolt!

Bill could not have known what size wrenches to bring unless he was the one who had *loosened* the nut and bolt.

Mary finished last in the one-miler.

But she had helped her teammates win the twelve-hour race. Their prize was a free trip to Washington, D.C., and the honor of mowing the south lawn of the White House.

Bill was banned from the mower races for five years.

SOLUTION TO
The Case of the Jumping Frogs

The thief was Hal, the tractor driver.

He had cut a firebreak, leaving a trail of ground-up bushes, rocks, and stumps. Then he had come back and stolen Ribbet's camera while everyone was frog hunting.

Hal had told the ranger he hadn't driven the tractor into the campgrounds again.

Encyclopedia proved he had.

After the frog hunt, the children had returned to the campgrounds and played leapfrog among the *tire tracks* in the firebreak.

Encyclopedia reasoned that the driver had lowered the wheels and raised the disks in order to travel quietly. The tire tracks showed he had driven into the campgrounds *after* cutting the firebreak.

Thanks to Encyclopedia's sharp brain, Ribbet got his camera back. Hal was fired.

SOLUTION TO

The Case of the Toy Locomotive

Encyclopedia saw through Sledge's lie.

A revolving door turns counterclockwise. So Sledge would have had his right arm to the open side of his door.

But Sledge had his right arm in a sling. The bag was over his left shoulder.

No thief would risk taking the time to squeeze into the revolving door with Sledge, reach around him, pull down his left arm, which was pushing the door, yank the bag free, and squeeze out again. Thieves would target someone carrying a bag over their right (or outside) shoulder.

Sledge confessed. The locomotive had not been stolen. He had hidden it, intending to come back for it. He had made up the theft so he could get another toy free.

Sledge kept the locomotive, which he had paid for, but gave back the music box. Sol bought the music box for five dollars and twenty cents and gave it to Birdie.

SOLUTION TO

The Case of the Air Guitar

One of the two boys overheard in the office said, "It's nearly twenty minutes past eleven." That's the way someone looking at a watch with hands states the time.

The other boy answered, "You're fast. It's only eleven-eighteen." That's the way someone with a watch that shows numbers digitally states the time. Then he added, "Oops, eleven-nineteen." That meant a number (the minute number, in this case) had changed, as numbers do on a digital watch.

The digital watch belonged to Phil Twining. He confessed. He and Manny Foster had dumped Herb's music down the sewer, hoping to better their chance of winning.

The judges disqualified them. They let Herb select another piece of music. He chose Scott's piece, "The Tasmanian Jump."

As he hadn't practiced to "The Tasmanian Jump," Herb finished second, behind Scott, who won.

SOLUTION TO

The Case of the Backwards Runner

In his eagerness to frame Felix as the thief, Rupert made a slip of the tongue. He said he saw Felix running down the exit lane after he had jogged *past* it.

That meant his back was to Felix. Sally caught the mistake.

So Rupert had to come up with an explanation fast. He said he was jogging backwards, and that was how he could see Felix.

"I always jog backwards when the sun is in my eyes, like today," he said.

That was his second mistake.

Encyclopedia knew the sun couldn't have been in Rupert's eyes, whether he had been jogging forwards or backwards.

It was noon, as the church bells proclaimed. Hence, the sun was directly overhead.

Rupert's father made him pay for the pen. The coach of the wrestling team made him run one mile after practice each day *backwards*.

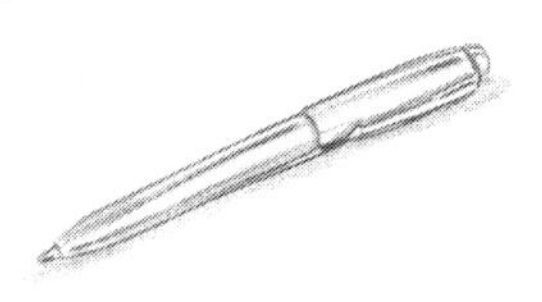

About the Author

Donald J. Sobol is the award-winning author of more than sixty-five books for young readers. He lives in Florida with his wife, Rose, who is also an author. They have three grown children. The Encyclopedia Brown books have been translated into fourteen languages.

DISCARDED